THE HISTORY AND D

OF

STEAM LOCOMOTION

ON

COMMON ROADS,

BY

WILLIAM FLETCHER,

MECHANICAL ENGINEER,

Author of "The Steam Jacket Practically Considered."

108 ILLUSTRATIONS.

LONDON: E. & F. N. SPON, 125, STRAND.
NEW YORK: 12, CORTLANDT STREET.
IPSWICH: PRINTED AND PUBLISHED BY S. H. COWELL.
1891.

All rights reserved.

Steam on Common Roads

Foreword

Although this title was reprinted some years ago by David and Charles, copies even of the reprint are extremely scarce. In furtherance of my desire to reproduce rare titles in small numbers, I have ordered an initial print run of 200 copies, which I hope will benefit transport historians and enthusiasts, to whom the task of accessing original copies might otherwise be difficult or very expensive. Please note that drawings have been reduced from the originals.

Adam Gordon, April 2006

ISBN: 1 874422 61 3

Republished by Adam Gordon, Kintradwell Farmhouse, Brora, Sutherland, KW9 6LU. Tel: 01408 622660

Publication no.63

Printed by 4edge Limited, 7a Eldon Way, Eldon Way Industrial Estate, Hockley, Essex, SS5 4AD

Covers prepared by Trevor Preece – email: Trevor@epic.gb.com

A catalogue entry for this book is available from the British Library.

PREFACE.

In dealing with Steam Transport on Common Roads the following division of the subject is usually adopted :—
1. Road Locomotives for the conveyance of Passengers, Parcels, and light goods at quick speeds.
2. Road Locomotives for Heavy Haulage and the conveyance of goods at slow speeds.
3. Agricultural Locomotives for use on farms.

In the present work the history of steam locomotion has been traced from the earliest down to modern times. The engines belonging to the first division being fully illustrated and described; the latter portion of the book also contains illustrations and much practical information relating to road locomotives placed under the second head.

The book is divided into seven sections :—
Introduction.
The Period of Speculation.
The Period of Experiment.
The Period of Successful Application.
The Modern Period.
Practical Notes on the Design and Construction of Road Locomotives.
Traction Engine Law.

The first five sections are chiefly historical, the Modern Period, however, contains many useful notes and tables of dimensions.

This work is a far more exhaustive treatise on this subject than has hitherto been attempted. There is not a book in existence which gives anything like a complete history of steam locomotion on common roads. In the two or three books (now out of print) treating on this and kindred subjects, the historical notes are most meagre and incomplete, only a few of the pioneers are mentioned, and some of those men who have done the most to further the progress of steam locomotion on the highway have received the least recognition.

Some individuals question the utility of engineering history, but to others this class of literature always forms an interesting and profitable study. Alderman W. H. Bailey, when recently addressing the engineering students of the Manchester Technical School, recommended the study of history as being the quickest and surest method of getting at the wisdom of the past. Mr. Bailey went on to say: "What is this historical knowledge but the rich record of the work of the best and cleverest and most successful men who have worked in the same direction as that in which you may be engaged? You can examine their difficulties, and appreciate their obstacles, and measure their mistakes and errors of judgment, and not only study their best, but also their worst work; and see the wise course which would have been successful, and sometimes why carelessness or want of judgment led to failure."

Amid the bustle and confusion of these high-pressure times, we, who reap the innumerable benefits which have been dearly bought and handed down to us by our predecessors, seldom stop to think of, much less to tender our thanks to those who have rendered such service in the past; and it is well that their worthy names should be chronicled and brought under our notice, and the memory of their deeds not allowed to die.

We not only overlook the work of our predecessors, but some individuals are in the habit of attributing some of the inventions of the less known engineers to those who are placed on the highest pedestals; for instance, quite recently some of Murdock's inventions were attributed to the illustrious Watt;

and now we have another speaker awarding honour to Symington which is justly due to Murdock. Sir Wm. Thompson, in his address at the unveiling of the Symington Memorial Bust, at Edinburgh, in November, 1890, places Symington as the first to apply steam power to the navigation of vessels. He went further, and claimed for him, in the model which Symington exhibited in Edinburgh in 1786, priority also in the application of steam power to the propulsion of road or rail carriages—in other words, as the author of the locomotive. It is well known that Murdock made the first model of a steam road locomotive in England several years before Symington constructed his model. In this book, we have placed both the above locomotionists in their proper order.

No apology need be offered for the publication of the "History of Steam Locomotion on Common Roads," seeing that no such history has heretofore been issued.

Some of the earlier portions of the book have appeared in the pages of "Industries," "The Mechanical World," and "The Practical Engineer."

The author is indebted to the proprietors of "The Engineer," "Engineering," "The Graphic," "Industries," "The Practical Engineer," and "The Mechanical World," for many of the illustrations.

The writer acknowledges with thanks some blocks kindly lent by Messrs. Aveling and Porter, of Rochester ; Messrs. Burrell and Sons, of Thetford ; Messrs. Foden and Sons, of Sandbach; Messrs. J. Fowler, Messrs. Hathorn Davey and Co., Messrs. J. and H. McLaren, of Leeds ; Messrs. Robey and Co., of Lincoln ; Messrs. Tangye, of Birmingham.

He wishes to acknowledge his indebtedness to Messrs. Hornsby and Sons, of Grantham, for drawing of road locomotive ; and also for valuable information placed at his disposal by Mr. Robert Edwards, of Grantham.

PREFACE.

The following books among many others have been of assistance to the writer :—

Historical Anecdotes of the Steam Engine, 2 Vols.	Stuart	1829
Historical Treatise on Elemental Locomotion	Gordon	1832
Narrative of Experiments with Steam Carriages	Hancock	1838
Steam Power applied on Common Roads	Maceroni	1835
Memoirs of Colonel Maceroni, 2 Vols.		1836
The Steam Engine—a lecture by E. A. Cowper		1884
Report on the Trials of Traction Engines		1871
Memoirs of Richard L. Edgeworth, 2 Vols.		1820
The life of W. Murdock	Buckle	1850
Description of Cugnot's Locomotive	Cowper	1853
Rise and Progress of Steam Locomotion on Roads	Head	1873
On the working of Traction Engines in India	Crompton	1879
Aids to Locomotion (abridgments of Patents)		1858
Economy of Steam on Common Roads	Young	1860
Men of Invention and Industry	Smiles	1884
Industrial Biography	Smiles	1879
James Nasmyth—an Autobiography	Smiles	1885
Life of R. Trevithick, 2 Vols.	Trevithick	1872
The Life of James Watt	Muirhead	1858
History and Progress of the Steam Engine	Galloway	1830
History of the growth of the Steam Engine	Thurston	1879
"One and All," Autobiography of Richard Tangye		1889
On Steam Carriages	Yarrow	1863
Steam on Common Roads	McLaren	1890
Hints to Purchasers of Traction Engines	Burrell	1890
Saturday Magazine		1839

Herbert's Engineers' Encyclopædia, 2 Vols.
'The Mechanic's Magazine," first 12 Vols.
" The Engineer."
" Engineering."

This book has been written during the leisure time in the intervals of business, and doubtless some of the faults of the work are occasioned by this circumstance.

W. FLETCHER.

HARLAND HOUSE,
   IPSWICH,
      December, 1890.

CONTENTS.

INTRODUCTION.

Cugnot. Murdock. Symington as successful promoters of locomotion. Watt's crude locomotive. Trevithick's simple steam coaches. Nasmyth's little carriage. Railway locomotion sixty years ago. Redivivus, on steam carriages. Abortive carriages not made by engineers. Good designs. Bad Workmanship. Constructive difficulties. Public opinion against them. Early carriages opposed by road authorities. Watt opposed steam locomotion. Dance's and Russell's carriages opposed. Geo. Stephenson on road locomotion. Foolish Acts of Parliament retarded progress. Gurney's carriages. Editor of " Mechanic's Magazine " on Gurney's 'perfect' carriages. " Glasgow Chronicle " on Gurney's break-downs. Hancock's success. Maceroni. Russell's and Hill's carriages. Sir James Anderson's road locomotives. Rickett's carriages. Schmidt's compensating gear. Adamson's road locomotive. Lough and Messenger's engine. Carrett's steering gear. Hodges. Richard Tangye's remarks. Thompson and rubber tyres. Bridges Adams's spring wheels. Aveling and Porter.

Pages 1 to 7.

THE PERIOD OF SPECULATION.
A.D. 1300, to A.D. 1769.

Early speculations. Roger Bacon. Dr. Wilkins on flying chariots; conveyances to the moon; 'castles in the air.' Archytas. Montanus. Dr. Darwin's prophecy. Steam balloons. Road Locomotion. Ramsey's engines for drawing carts. Solomon De Caus. Sir Isaac Newton's prophecy; his road locomotive; an idea for others to work out. Father Verbiest; his experiments at Pekin. Papins belief; use of ratchets. Leupold's scheme. Savery. Dr. Robinson; steam locomotive proposals. Watt and Robinson. Darwin. Boulton's fire engine. Franklin. Fiery chariot. Darwin's equipage. Darwin and Boulton. Darwin suggests a steam carriage; requests Boulton to become a partner. Moore's locomotive. "Leed's Mercury" on Moore's invention; three patents; sells his horses. Dr. Small. Watt on the folly of inventing; threatens to stop Moore.

CONTENTS

THE PERIOD OF SPECULATION—*continued*.

Dr. Small's rotary engine. Edgeworth's portable railway; sailing carriages; one hundred models. Boydell's endless railway. Edgeworth and Darwin. Edgeworth and Watt.

Pages 9 to 22.

THE PERIOD OF EXPERIMENT.
A.D. 1770 to A.D. 1831.

Cugnot born: makes a steam carriage. Planta's model. Cugnot's locomotive: makes a larger one; his high pressure engine, Cugnot's pension and death. Murdock; his early years; leaves Scotland for Soho; his interview with Boulton: the wood hat; enters the Soho firm. Murdock's locomotive experiments: frightens the vicar of Redruth. Mr. R. Tangye. Watt's displeasure. Symington and Sadler. Murdock's accident and death. Watt; opposes road locomotion; his patent specification; wood boiler; crude ideas; spur gearing; promises to make a model. Dr. Black. Symington; makes a model road engine; exhibits it at Edinburgh; abandons the subject. Millar. Symington's engine for boat. Lord Dundas. Symington's patent. The 'Charlotte Dundas.' Symington's poverty, annuity and death. Sadler; makes experiments; his rotary engine. Evan's ideas on locomotion: his difficulties; builds an engine; changes his plans; makes mill engines. 'Oructor Amphibolis.' Fourness and Ashworth's steam carriage. Allen. Read's vertical boiler; his steam carriage; makes a model. Trevithick; father of the high pressure engine. First road locomotive; the Camborne locomotive; the boiler too small. Trevithick and Vivian became partners; their London steam carriage; wrought-iron boiler; spur gearing: trial trips and mishap. Trevithick's railway engine experiments; his last days at Dartford, and death. Dumbell. Pratt. Stevens. Palmer. Tyndall. Brunton's 'mechanical traveller'; propellers. Reynolds; patents a locomotive. Gordon's gas apparatus; gives attention to road locomotion. Large driving wheels. Gordon patents a steam carriage; propellers; his many experiments. Griffiths patents a carriage. Made by Bramah. Brown. Burstall and Hill's carriage; all the wheels used as drivers; ratchet wheel and spring pawl. Second patent carriage; exhibited at Leith. James's locomotive; separate from the carriage. James and Anderson; steam carriage trials. Neville's steam carriage; spring wheels. Seaward. Parker. Andrews inventor of the 'pilot' steering wheel; used by Gurney and Aveling. Gough. Holland. Nasmyth; makes a carriage; successful test; used the exhaust steam to create a draught. Viney. Harland patents a steam carriage; elegant form; his love of mechanical pursuits. Sir Geo. Cayley. Harland mayor of Scarborough. Rawe and Boase. Clive. "Saxula." Lea. Heaton. Heaton's steam carriage company. Napier.

Pages 23 to 94.

CONTENTS.

THE PERIOD OF SUCCESSFUL APPLICATION.
A.D. 1831 to A.D. 1867.

Summers and Ogle's steam carriage; extraordinary speeds. Gurney occupies a prominent place; his lectures; constructs a locomotive; used propellers. Sir Chas. Dance. Gurney's coaches run between Gloucester and Cheltenham; opposed by the public; prohibitory turnpike rates. Ward at Glasgow. Gurney's 'perfect'; steam drags; he petitions the House of Commons. Sir Geo. Cayley supports the petition. Sir Goldswothy Gurney's later inventions; Bude light; his death. Sir Chas. Dance. Dr. Church's novelties; beautiful coach. Church built the 'Eclipse' railway locomotive. Yates and Smith. Field's improved boiler; one of the founders of the Inst. of Civil Engineers. Millichap. Hancock invented a good boiler; built a number of steam carriages; the 'Infant'; the 'Autopsy'; the 'Enterprise' The Era was shipped to Dublin. Hancock ran carriages through the London streets; steam omnibus opposed; the London and Paddington steam carriage Co; Redmund's dishonesty. Hancock's last steam carriage; Automaton ran 15 miles an hour; built ten carriages, in 16 years, all well made. Redmund's steam carriage; a copy of Hancock's; a complete failure. Maceroni's early years; Aide-de-camp to King of Naples; helping Gurney: entered into partnership with Squire: they patent a boiler; make a good carriage; run many trial trips. Bushy Heath. Partership dissolved. A carriage sent to France. Maceroni lost two carriages; he was in great distress. Beale builds a third carriage. Deitz. Gibbs and Applegarth complete a steam carriage; a curious boiler used. Watts. Roberts's steam locomotive; ran 20 miles an hour; met with an accident. Inventor of the compensating gear. Carrett Marshall. Roberts's poverty and death. Russell's early life; the designer of the 'Great Eastern'; patents a locomotive; six coaches built to his designs; his coaches were popular; an experienced engineer. Multitubular boiler. Road authorities put coating of loose stones on the road. An axle broke. Russell sends two coaches to London. Hill takes a lesson in steam carriage construction; makes a carriage; uses the compensating gear. Ran a carriage 128 miles in a day. Sir James Anderson engaged in steam carriage construction; a company launched to run his carriages. Anderson's drags built in Manchester and Dublin; purposed great things; little was accomplished; he devoted 31 years of his life to the furtherance of locomotion. Squire. Norrberg. Bourne. Fisher; his steering gear. Dudgeon. Rickett makes an engine for the Marquis of Stafford; builds two more engines, one sold to Earl of Caithness; who rode over one of the largest and steepest hills in Scotland. Rickett sends an engine to Spain. Seaward. Adamson made an engine for Schmidt. Race between Adamson's and Boulton's steam carriages. Lough and Messenger. Bach. Stirling. Carrett builds an engine for Mr. Salt; a noted carriage; ran 15 miles an hour. Mr. Hodge's Fly-by-night was often in trouble. Smith. Yarrow and Hilditch, radius link. Lee. Hayball. Wilkinson. Tangye's carriage. Boulton. Goodman. Armstrong. Pages 95 to 170.

B

x.

CONTENTS.

THE MODERN PERIOD.
A.D. 1868 to A.D. 1891.

Thompson's early life, road steamers, indiarubber tyres, chain armour, advantages of indiarubber tyres, Wolverhampton show, great demand for Thompson's engines. Tennant. Todd. Fisher's parallel rods. Nairn's omnibus, rope tyre wheel, the "Pioneer." Knight. Catley. Robey's engines. "Advance," gearing on Thompson's engines, test of the "Advance." Thompson's "Pot" boiler. Ransome's first road locomotive, first ploughing engine, Kilburn show 1879, Ransomes make Thompson's road steamers for Indian Government, ran from Ipswich to Wolverhampton, "Chenab" and omnibus, Mr. Crompton in charge of the "Chenab," "Ravee" ran from Ipswich to Edinburgh and back. Mr. Crompton's paper 1879, no limits to speed in India. Mr Head's paper 1873. British engine drivers, native drivers do well. Mr. Muirhead. Ransomes' traction engines. Burrell's first traction engine, Boydell's endless railway, good workmanship, Burrells make Thompson's road steamers, three sent to Turkey, engine at the Wolverhampton show, road steamers with horizontal boilers, engine and large omnibus made by Burrell, interesting trials at Thetford. chain armour, improved shoes, road locomotive on springs, spring mounted traction engines, single crank compound engine, Windsor show 1889. Aveling's early engines, chain traction engines, road engine "El Buey," Oxford show 1870, geared engines introduced, Aveling and Greig's rubber tyre wheel, "Steam Sappers," Bridges Adams's wheels, Paris Exhibition engine 1878, inside gear engines, crankshaft brackets, road locomotive 1890, road locomotive and dynamo. Fowler's ploughing machinery, Wolverhampton show, compound road loco. on springs, dimensions of engine. Hornsby's first road engine, road locomotives, Burrell and Edwards' clutch gear, gearing dimensions, speed ratios. Mackenzie. Perkin's make a novel compound road locomotive, London Exhibition 1873, something like Cugnot's, economical, noiseless, used by Yorkshire Engine Co., several tests carried out. Archer and Hall inventors of a road locomotive. McLaren latest and best road engines, one sent to India, compound on spring wheels, interesting test near Leeds, McLaren's omnibus for India, send three fine engines to France, Fourgon poste, parcel service, engines run in the night, dangerous roads, steam brake, gas head light, engine weighs 15 tons. McLaren and Boulton's patent wheel, used by Fowler, Aveling, and others, wood blocks bedded on felt, noiseless, do not slip, no damage done to the roads. Foden's first traction engine, possessed wide departures, double cylinder traction engines, advantages of double cylinder engines, large driving wheels, piston valves, balanced valves, all engines mounted on springs, compound cylinder, auxiliary valve, water heater, Stockport trials, Newcastle engines, advantages of compounding. Concluding remarks. All high speed engines for foreign countries. Horses soon reconciled to the sight of traction engines. End of historical part.

Pages 171 to 257.

CONTENTS.

PRACTICAL NOTES ON THE DESIGN AND CONSTRUCTION OF ROAD LOCOMOTIVES.

Table of dimensions of simple and compound road locomotives, cylinders, steam jacketed, large ports, steam dome, compound road locomotives, proportions of cylinder areas. Boilers, thickness of plates, heating surface and grate area per horse power. Box brackets, stiffening plates, steel carriages. Tender and tank-drawbar, spring coupling. Driving wheels, how constructed, two types, brake barrel. Fore-carriage, steerage gear, box for spuds on front axle. Compensating gear, details of, method of locking the compensating gear, gradient 1 in 11 wheels slipping. Winding drum, wire rope, how used, winding drum and compensating gear combined, guide rollers fixed at the bottom. Steel gearing, fast and slow speed ratios, Hornsby's pinions. Number of countershafts, three shaft engines, four shaft engines, advantages of each type. Crankshafts, solid keys, eccentrics forged on. Phosphor bronze bearings. Connecting rod, two bolts through strap, set pin. Link motion, slip of the die, equal cut-off at each end, large pins, lubrication of slide valves, cast iron eccentric straps, flat eccentric rods. Feed pumps, fixed on boiler barrel, long delivery pipes, solid plunger, ball valves, round flanges on pipes, water heater, injector, water lifter and hose on fore tank. Few holes in boiler, reduce number of details, water-filling holes. Governors, direct acting, eqilibrium throttle valves. Stop valve gear, displacement lubricators. Bored guides. Spring balance safety valves, Ramsbottom type valves. Handy arrangement of levers, handles, cocks, &c. Damper in chimney, joint on chimney. Weight on driving wheels, weight on leading wheels. Cost of haulage. McLaren's engines, table of effect on roads, &c. Lincoln traction engine table. Beauty of design, ribbed castings ugly, graceful forms to be aimed at, drawing office care, simplicity, efficiency not be sacrificed, improvement in design, *The Engineer* on traction engines, the design an index to the character of the engine.

Pages 259 to 278.

TRACTION ENGINE LAW.

Legal restrictions, engines termed a nuisance, a license to be obtained, four miles an hour maximum speed allowed, man to walk in front, not allowed to cross bridges, crusade against traction engines, not to blow off steam, foolish regulation, to consume the smoke is impossible, regulations as to crossing bridges, Aveling's remarks on this subject, engines have to run in the night, accidents occasioned by this legislative enactment, Maidstone boiler explosion, damage to bad roads, spring engines, Sheffield memorial, meeting respecting narrow roads.

Pages 279 to 288.

CONTENTS.

LIST OF TABLES.

DESCRIPTION.	Page.
Record of the Trip of the Ravee to Edinburgh and back	196
Particulars of Fowler's Compound Locomotive	228
Dimensions of Hornsby's Gearing for Road Engine	233
Dimensions of Foden's Road Locomotives at Newcastle	254
Analysis of Waste Gases of best Engines tried at Newcastle	256
Dimensions of Simple and Compound Road Locomotives	260
Nett effect of McLaren's Engines on the road	275
Dimensions of Single and Double Traction Engines by a well-known Lincoln Firm	276

DIRECTORY OF LEADING MAKERS OF ROAD LOCOMOTIVES AND TRACTION ENGINES.

W. Allchin.

Chas. Burrell & Sons.

Richard Hornsby & Sons.

Ransomes, Sims & Jefferies.

Aveling & Porter.

E. Foden & Sons.

J. & H. McLaren.

Robey & Co.

Pages 291 to 303.

LIST OF ILLUSTRATIONS.

Figure.		Page
1.	Sir Isaac Newton's Locomotive, 1680	13
2.	Cugnot's Steam Carriage, Side view, 1770	26
3.	Cugnot's Steam Carriage, Plan	26
4.	Murdock's Model Locomotive, 1781	31
5.	Watt's Gearing for Road Locomotive, 1784	26
6.	Symington's Steam Coach, 1786	40
7.	Read's Steam Carriage, 1790	47
8.	Trevithick's Model Locomotive, Side view, 1796	50
9.	Trevithick's Model Locomotive, End view	50
10.	Trevithick's Road Locomotive, Side view, 1801	53
11.	Trevithick's Road Locomotive, Plan	53
12.	Trevithick's Return Flue Boiler, 1801	54
13.	Trevithick's Steam Coach, Side view, 1803	56
14.	Trevithick's Steam Coach, Plan	56
15.	Brunton's Locomotive, 1813	60
16.	Gordon's Road Locomotive, 1821	63
17.	Gordon's Steam Coach, 1824	64
18.	Gordon's Propellers, 1824	65
19.	Griffith's Steam Coach, 1821	67
20.	Burstall's Steam Coach, 1824	69
21.	Burstall's Gearing	71
22.	Burstall's Ratchet Wheel	71
23.	Burstall's Steam Coach, 1827	73
24.	James's Engines and Boiler, Plan, 1824	75
25.	James's Engines and Boiler, End view	76
26.	James's Engine, Side view	77
27.	James's Steam Coach, 1829	78
28.	James's Steam Locomotive, 1832	79
29.	Neville's Spring Wheel, 1827	83
30.	Neville's Steam Carriage	84
31.	Seaward's Propeller, 1825	85
32.	Andrew's Pilot Wheel, 1826	86
33.	Andrew's Section of Boiler	87
34.	Napier's Steam Carriage, 1831	93

LIST OF ILLUSTRATIONS—*continued*.

Figure.		Page
35.	Summer's Steam Carriage, 1831	95
36.	Gurney's Steam Carriage, 1828	99
37.	Steam *versus* Horses	100
38.	Gurney's Steam Drag, Side view, 1830	101
39.	Gurney's Steam Drag, Plan	101
40.	Church's Steam Coach, 1835	106
41.	Hancock's Steam Carriage, 1827	110
42.	Hancock's "Infant," 1831	111
43.	Hancock's "Enterprise," 1833	112
44.	Hancock's "Automaton"	117
45.	Hancock's Steam Gig, 1838	115
46.	Redmund's Steam Carriage	120
47.	Maceroni's Steam Carriage, 1833	124
48.	Russell's Steam Carriage, 1834	133
49.	Russell's Steam Engine	134
50.	Early Compensating Gear	138
51.	Modern Compensating Gear, 1878	139
52.	Hill's Steam Carriage, 1841	141
53.	Fisher's Steering Gear, 1853	148
54.	Rickett's Steam Carriage, 1858	150
55.	Rickett's Road Locomotive and Coach, 1865	153
56.	Lough and Messenger's Carriage, 1858	156
57.	Carrett's Road Locomotive, 1862, (Folding leaf)	158
58.	Carrett's Steering Fork	159
59.	Carrett's Improved Steerage, Side view	160
60.	Carrett's Improved Steerage, Plan	160
61.	Yarrow's Steam Carriage, 1862	162
62.	Tangyes' Road Locomotive, 1862	166
63.	The "Cornubia" on the Village Green	168
64.	Thompson's Road Steamer, 1868	173
65.	Thompson's Rubber Tyre Wheel, 1868.	175
66.	Thompson's Wheel and Armour, Side view	177
67.	Thompson's Wheel and Armour, Section	177
68.	Creep of Rubber Tyre	178
69.	Nairn's Road Wheel	183
70.	Robey's Road Steamer "Advance," 1870	186
71.	Robey's Sectional View of "Advance," (Folding leaf	188
72.	Robey's Gearing, Side view	187
73.	Robey's Gearing, End view	187
74.	Thompson's "Pot" Boiler, made by Robey, 1870	191
75.	Ransomes' Road Steamer and Tender (Folding leaf)	196

LIST OF ILLUSTRATIONS—*continued*.

Figure.		Page.
76.	Ransomes' Traction Engine	199
77.	Burrell's Road Steamer, 1870	201
78.	Burrell's Road Locomotive, 1871	203
79.	Burrell's Road Locomotive and Coach	204
80.	Burrell's Road Locomotive (Folding leaf)	204
81.	Burrell's Rubber Tyre and Armour, Section	205
82.	Burrell's Rubber Tyre and Armour, Side View	206
83.	Burrell's Patent Spring Mounted Road Engine, 1890	208
84.	Burrell's Spring Mounted Traction Engine (Folding leaf)	209
85.	Burrell's Spring Mounted Traction Engine ,,	209
86.	Burrell's Spring Mounted Traction Engine ,,	209
87.	Burrell's Compound Cylinder, 1890	210
88.	Aveling and Greig's Rubber Tyre Wheel, Side View	214
89.	Aveling and Greig's Rubber Tyre Wheel, Section	214
90.	Bridges Adams's Wheel, Section	215
91.	Bridges Adams's Wheel	216
92.	Aveling's Inside Gearing, Plan, 1878	218
93.	Aveling's Inside Gearing, Section	219
94.	Aveling's Inside Gearing, Plan, 1889	220
95.	Aveling's Road Locomotive, 1890	222
96.	Aveling's Road Locomotive and Dynamo, 1890	223
97.	Fowler's Compound Road Locomotive, 1890	226
98.	Hornsby's Compound Road Locomotive, 1890 (Folding leaf)	230
99.	Burrell and Edwards' Clutch Gear, Side View	231
100.	Burrell and Edwards' Clutch Gear, Plan	231
101.	McLaren's Indian Road Locomotive, (Folding leaf)	236
102.	McLaren's Indian Omnibus, 1885	239
103.	McLaren's French Road Locomotive, 1886 (Folding leaf)	240
104.	McLaren and Boulton's Wheel, 1891	243
105.	Foden's Road Locomotive, 1890	247
106.	Foden's Spring Mounted Road Engine, 1891	249
107.	Foden's Compound Cylinder, Section	251
108.	Allchin's Traction Engine, 1891	258

INTRODUCTION.

The very early history of steam locomotion on common roads during the periods of speculation and experiment is little more than a recital of failures. Cugnot in France, and Murdock and Symington in England were the first promoters who attained any measure of success, the two latter inventors, however, only produced models of their road locomotives. Several other patentees, including the illustrious Watt, about this time turned their attention to this subject with indifferent results. Watt's chief anxiety appeared to be to prevent other people from constructing steam carriages, his own ideas were very crude; he purposed using a low pressure wood boiler, the sun and planet motion, and his separate condenser, on his road locomotive. As soon as Trevithick approached the steam carriage enterprise, rapid progress was at once apparent, he was the first engineer who conveyed passengers by steam on an English highway. This celebrated genius carried into effect the proposals and speculations of many of his predecessors; his locomotives were models of simplicity, and had Trevithick received any pecuniary aid greater success would have been achieved by him. He ceased to build and run steam carriages because of the great expense it entailed. Passing over many names whose works do not call for any remarks, we reach the date of Nasmyth's little experimental engine which was made for the members of the Society of Arts in Scotland. It is to be regretted that no particulars are furnished us respecting the design and constructive details of this interesting engine. The small sketch of the engine given in the Autobiography does not help us as it shews no details clearly.

Sixty years ago (the date of the commencement of the period of successful application) railway locomotion was attracting a good deal of attention, and road locomotion was at a very low ebb. Notices of steam road carriage experiments became less numerous in the journals and were set up in small type, while pamphlets were hurried from the press urging the construction of railways, and the newspapers of the period were crowded with long articles setting forth the progress made in the newly developed mode of transport. Moreover, some of the road locomotionists had brought an amount of discredit upon their cause, by foolishly attempting to rival the railway trains, their extravagancies tending to retard progress in road locomotion. The one aim of the carriage proprietors appeared to be to outrun each other, whereas had they been satisfied with a reasonable speed—say of eight or ten miles an hour—instead of boasting of having run twenty miles an hour; and had they regarded their steam road conveyances as feeders to the great railway system, good would have resulted from their introduction in many districts. Redivivus, who wrote much on steam carriages before and after the introduction of railways said:—" There can be no rivalry between railways and highways; each have their appointed purpose to fulfil. Railways carry cheaper and faster, but there is plenty of scope for the slower speed of the road locomotives. Let a company be formed and offer a premium for the best road engine, and when the best is chosen let the company have one made at their expense, in the best style, because, he says, " the poor inventors have to make shifts and use improper materials because of their lack of funds—are apt to spoil the ship for a hap'orth of tar."

INTRODUCTION.

Some of the steam carriage promoters possessed little or no knowledge of the requirements of such engines, and had received no practical acquaintance with machinery, and, as might be anticipated, in attempting to solve such a difficult engineering problem they miserably failed. Several abortions of such schemes which we could name were brought out for a brief airing, and soon displayed some irritating traits of character on the highway, not unfrequently ending the performance very abruptly with a break-down. The steam carriage exploits were invariably witnessed by a member of the press, who had by invitation accompanied the sanguine inventor during the trip. The account of the trial formed a theme for a commendable article in the daily paper. Need we add that the writer was,

> To their virtues very kind,
> To their faults a little blind.

These carriages, after acting their part in running one or two short experimental trips on a level and smooth road were mysteriously disposed of; so they and their novel manœuvres were speedily forgotten. Some of the early locomotionists produced a fairly good plan, in the working out of which considerable ingenuity and mechanical skill were displayed, but good workmanship could not be obtained; consequently during each trial some hitch occurred traceable to defective workmanship. But most of the steam carriage projectors were not to be turned aside by trifling breakages, and by the application of an indomitable perseverance, they, out of repeated failures, eventually achieved a mechanical success.

However, after surmounting constructive difficulties, and in possession of an engine that worked fairly well, they were met by a still more formidable obstacle, to which they one after another succumbed.

INTRODUCTION.

Public opinion was against them; the influential people who opposed railways about this time, but in vain, did their utmost to put a stop to steam locomotion on the highway, and in many cases succeeded in their design.

The steam carriages of the early days were hooted and hissed at by road authorities, coach-drivers, publicans, and the multifarious men who delighted in horses, which perhaps was not to be wondered at, seeing that the interests of these men were at stake; but engineers and others who ought to have known better, joined in the crusade against steam on common roads, from the celebrated James Watt, who actually put a covenant in the lease of his house, Heathfield Hall, that "no steam carriage should on any pretence be allowed to approach the house."*

He also discouraged Murdock when busy with steam carriage schemes. Could his prejudice against steam locomotion have gone any further? Such were the facts, and the antagonistic spirit displayed by Watt, when penning the foolish provision respecting his house, is in keeping with the feelings evinced by succeeding generations of steam locomotion obstructionists, who placed loose stones 18 inches deep on the Cheltenham and Gloucester Road, for the purpose of disabling Sir Charles Dance's steam carriages, and thus preventing their running. The same determined opposition, accompanied by similar treatment, was extended to Scott Russell's carriages on the Paisley and Glasgow road; and an Act of Parliament was speedily passed to levy prohibitive tolls on steam carriages.

Some engineers who did not actually oppose steam locomotion on common roads were evidently prejudiced against its introduction. Geo. Stephenson for instance, said "steam carriages on ordinary roads would never be effective, or at least sufficiently serviceable to supersede horse carriages;" he

* "The Steam Engine" by Mr. E. A. Cowper, M.Inst.C.E. 1884.

had, however, paid no attention to the question and was not an authority on this subject, consequently his opinion was at fault.

Contemporary engineers knew that steam locomotion on common roads at quick or slow speeds was practicable. And had not foolish laws stepped in their way, we should have had many instances of quick travelling on our highways.

Bearing in mind the combined forces that have barricaded the path of steam power on roads during its history from the earliest times, we cannot be astonished that its growth and development have been so slow.

Among the carriage proprietors of the period of successful application was Gurney, whose works have received more praise than is their due. Dr. Lardner placed Gurney at the head of the list of steam carriage inventors, and other authors followed in his footsteps. During the time that Gurney was petitioning the House of Commons and the committee recommended a grant of £16,000, the Editor of "The Mechanics Magazine" said :—"Never was there a person who had less claim on the national purse. He has left steam locomotion where he found it." His so called "perfect" steam carriages had to stop we are told every few miles for repairs.

The "Glasgow Chronicle" said : — "Gurney brought a steam carriage to Glasgow, but instead of coming by road through some great English towns he brought it per smack to Leith, and when he tried to travel from Edinburgh to Glasgow he required horses to help him up the hills."

Undoubtedly Hancock was the most successful of all the steam carriage builders thus far noticed. A writer in 1835 said :—"Gurney, Dance, Ogle, Maceroni, Russell, &c. have all talked much of their success, and of the wonders they have performed, Hancock has boasted none, and has accomplished a great deal ; the slow motion which he applied to his steam coaches was a great improvement, which every practical mechanic appreciated."

Maceroni was a clever carriage builder, he was particularly unfortunate in his business, and took great delight in ridiculing the work of his contemporaries.

While Gurney's and Hancock's carriages have received a great amount of attention from writers on this subject, Russell's, Hill's and Maceroni's have been nearly overlooked, we have tried to supply the missing links in this work.

If Sir James Anderson had one of his powerful carriages built both at Manchester and Glasgow as stated in the journals of the period, it is surprising that so little is said respecting them after completion. We have undoubtedly some citizens in our midst who could tell us what became of Russell's and Anderson's steam coaches, and these persons could furnish us with some interesting reminiscences of steam carriage exploits of 50 years ago.

After the disappearance of Anderson's carriages the enterprise met with a sudden and complete check, and it appeared as though the road-menders had managed to stamp the locomotives out of existence. From 1840 to 1857 no attention was paid to quick speed road engines until Rickett revived the subject, and brought out a number of good locomotives after 1857.

Messrs D. Adamson and Co. made a road locomotive for Mr. Schmidt, who patented the compensating gear as late as 1868, when the arrangement was introduced many years before by Mr. Roberts, used by Hill in 1834, and Carrett in 1862.

Lough and Messenger's small engine was little better than a steam tricycle, but a successful machine of small weight.

Mr. Carrett's steering gear was a decided improvement and quickly adopted by subsequent makers. This engine was practically stopped by the one-sided laws in force, which prevented Hodges from running the machine at anything higher than a crawling speed. The happy possessor of this locomotive had six summonses in six weeks.

Mr. Richard Tangye says respecting their road locomotive; "preparations were being made for doing a considerable business in these engines, but the 'wisdom' of Parliament made it impossible. Thus was the trade in quick speed road locomotion strangled in its cradle."

In the Modern Period we have dealt fully with the india rubber tyres introduced in 1867 by Mr. Thompson which marked the dawn of a new era in steam locomotion. We may say that no spring system was so effective as the rubber tyres, and now that we can make lighter engines, there can be no hindrance to the adoption of india-rubber in the form of tyres, or applied in other ways so as to make perfect spring engines. Mr. Bridges Adams's wheel as used so long by Messrs Aveling and Porter was a very good arrangement. The great cost of the rubber tyres was, we believe, the chief objection to their universal adoption. If the rubber manufacturers can produce these tyres or segments at a cheaper rate, we may possibly return to their use before long.

THE PERIOD OF SPECULATION.

IN studying the history of steam transport on the highway, we find speculations bearing upon the subject of greater antiquity than is generally supposed, and contemporary with these proposed steam carriages there are some schemes of rather unlikely modes of transport. Doubtless the proposals for carrying passengers by steam power, either on land or water or through the air were looked upon by the inhabitants of the period as

>—— "Speculations wild,
> And visionary theories absurd,
> Compared with which, the most erroneous flight
> That poet ever took when warm with wine,
> Was moderate conjecturing."

—The application of the force of steam for propulsion on sea and land was very possibly anticipated 600 years ago by Roger Bacon, the learned Franciscan monk, who, in the age of ignorance and intellectual torpor wrote:—" We will be able to construct machines which will propel large ships with greater speed than a whole garrison of rowers, and which will need only one pilot to direct them; we will be able to propel carriages with incredible speed without the assistance of any animal; and we will be able to make machines which by means of wings will enable us to fly into the air like birds."

While steam transport on land and water is an accomplished fact, yet steam flying chariots for the conveyance of passengers and goods through the air on the wings of the wind are as far from realisation to-day as ever they were. Dr. Wilkins, Bishop of Chester, whom Stuart describes as a person of "rare gifts" (and, judging by what follows, we do not doubt the statement), thought it not at all improbable for some of his posterity to find out a conveyance to the moon. He says :—" I do seriously, and upon good grounds, affirm it possible to make a flying chariot, in which a man may sit and give such a motion unto it as shall convey him through the air ; and this perhaps might be large enough to carry divers men at the same time, together with food for the journey. It is not the bigness of anything in this .kind that can hinder its motion, if the motive faculty be answerable thereunto. We see a great ship swim as well as a small cork ; and an eagle flies in the air as well as a gnat. This engine may be contrived from the same principle by which Archytas made a wooden dove, and Regio Montanus a wooden eagle. I conceive it were no difficult matter, if a man had leisure, to show more particularly the means of composing it. The perfecting of such an invention would be of excellent use, and would be of inconceivable advantage for travelling, above any other conveyance that is now in use, so that, notwithstanding all these seeming impossibilities, 'tis likely enough that there may be a means invented of journeying to the moon ; and how happy shall they be who are first successful in the attempt. Might not a " high pressure " be applied with advantage, to move wings as large as those of the " rucks " or the " chariot." The engineer might possibly find a corner that would do for a coal station near some of the *castles in the air.*"

Dr. Darwin, a century later, prophesied the eventful success of the steam flying chariot, as well as the locomotive and the steamboat, in the following often quoted lines :—

> Soon shall thy arm, unconquered steam, afar
> Drag the slow barge, or drive the rapid car ;
> On on, wide-waving wings expanded bear
> The flying chariot through the fields of air,
> Fair crews triumphant, leaning from above
> Shall wave their fluttering kerchiefs as they move,
> Or warrior bands alarm the gaping crowd
> And armies shrink beneath the shadowy cloud.

Since Darwin, in 1791, penned the above lines, thousands of minds have attempted to solve the problem of steam balloons. Numerous plans have been patented—some have been tried—but hitherto none have succeeded.

Leaving the visionary flying carriages, we shall now confine our attention to the daybreak of the less romantic and more matter-of-fact subject of steam locomotion on the highway.

RAMSEY.— In 1618 Ramsey and Wildgoose took out a patent for "newe, apte, or compendious formes or kinds of engines or instruments to ploughe grounds without horse or oxen ; and to make boates for the carryage of burthens and passengers runn upon the water as swifte in calmes, and more safe in stormes, than boats full sayled in great wynnes." It is probable that the patentees purposed ploughing by steam power, and, judging by what is stated in Ramsey's later inventions, we should imagine that steam engines were intended for the drawing of carts, for instance : " A farre more easie and better waye for soweing of corne and grayne, and alsoe for the carrying of coaches, carts, drayes, and other things goeing on wheels, than ever yet was used and discovered."

It is usual, in writing the history of the locomotive, to make Solomon De Caus talk to an imaginary marquis, of navigating ships, as well as of moving carriages, and of working other miracles by steam power. But this conversation has been proved by Muirhead to be a mere fable.

NEWTON.—It is curious that Sir Isaac Newton, in one of his books, said that it would be necessary that a new mode of travelling should be invented. He prophesied that the time would arrive when owing to the increase of knowledge, we should be able to travel at the rate of fifty miles an hour. These remarks were ridiculed at the time, but have been more than realised. Moreover, the world is indebted to the same illustrious personage for the first idea of propulsion on land by steam power, for in his "Explanations of the Newtonian Philosophy," written in 1680, he suggested the little locomotive shown by the accompanying engraving (Fig. 1), "which will be recognised as representing the scientific toy which is found in nearly every collection of illustrative philosophical apparatus." It consists of a spherical generator B; the driver sitting at A controls the escape of the steam by the lever E, and the cock F; the fire beneath the boiler is seen at D; the whole is mounted on light wheels, so as to move easily on a horizontal plane, and upon opening the cock F, steam would issue violently out of the nozzle C, shown pointing backwards, and by its reaction the carriage would be driven in the opposite direction, and propelled forward as indicated by the arrow. It would be very interesting to know whether Sir Isaac Newton ever made a model of his proposed locomotive; doubtless "he merely threw out the idea for other minds to work upon."

In Muirhead's "Life of Watt" it is mentioned that the Abbé Huc, in giving an account of Father Verbiest, a missionary among the Chinese, who died in 1688, says: "It is highly probable that he anticipated the great discovery of modern times, the motive power of steam." In his learned work entitled "Astronomia Europæa" there is a curious account of some experiments that he made at Pekin, with what we may call steam engines. "He placed an æolipile upon a car, and directed the steam generator within it upon a wheel to which four wings were attached; the motion thus produced was communicated by gearing to the wheel of the car. The machine continued to move with great velocity as long as the steam lasted; and, by means of a helm, it could be turned in various directions."

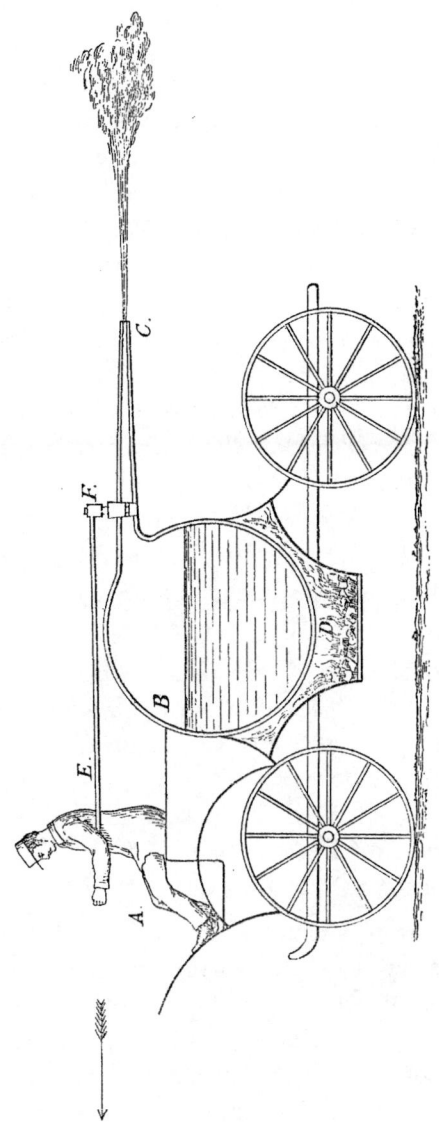

SIR ISAAC NEWTON'S STEAM ROAD LOCOMOTIVE, 1680. Fig. 1.

PAPIN.—In a letter of Papin's to Leibnitz, dated 25th July, 1698: " Believing as I do," says Papin, " that this invention—(the steam engine which he was then at work on)—may be applied to several other purposes than that of raising water, I have made a little model of a carriage which is driven by this force ; but I think that the inequalities and the curves of roads will make it very difficult to bring the invention to perfection."

Papin, in his work published at Capel in 1699, suggested the use of ratchets to convert the motion of a piston into a circular motion ; but it does not appear that he had any idea of a locomotive, although some of his successors have made use of ratchet gearing on their steam carriages, as we shall show in due course.

Leupold's scheme for a steam carriage, mentioned in "Theatrum Machinarum," was simply a passing thought, and probably the subject was never matured.

SAVERY.—Capt. Savery is credited with having given attention to the subject of steam locomotion, but we find that he merely hinted that his engine might be very useful in ships, but that he cared not to meddle with the matter, and left it to the judgment of those who were acquainted with such affairs. He has no claim to be classed among the inventors of steam locomotion.

Nearly a century elapsed after Sir Isaac Newton promulgated his steam carriage without any further progress being made in the subject ; but towards the end of the 18th century the steam engine was being rapidly applied to a number of useful purposes. Therefore, we need not wonder that many inventors turned their attention to the adaptation of steam power to the propulsion of carriages on common roads.

ROBINSON.—Dr. Robinson, in 1759, proposed the application of steam power for driving carriages to James Watt, who wrote as follows:—" My attention was first directed to the subject of steam engines by the late Dr. Robinson, then a student in the University of Glasgow, afterwards Professor of Natural Philosophy in the University of Edinburgh. He, in 1759, threw out the idea of applying the power of the steam engine to the moving of wheel carriages, and to other purposes, but the scheme was soon abandoned on his going abroad."

It has been truly stated that Watt, at the time of the above conversation, was more ignorant than his companion of the principles which were involved in the construction of the steam engine, and this suggestion of Robinson's may have had some influence in determining Watt to pursue his research; thus setting in operation that train of thoughtful investigation and experiment, which finally earned for him his splendid fame. Although Watt was giving his time and attention to the study of the steam engine, yet he allowed the project of steam locomotion to slumber for twenty-five years before taking any action in the matter.

DARWIN.— Six years after Dr. Robinson's proposal, Benjamin Franklin (then Agent in London for the United Provinces of America), Matthew Boulton, of Birmingham (subsequently Watt's partner), and Dr. Erasmus Darwin, of Lichfield, were in correspondence relative to steam as a motive power. It appears that Boulton had made a model of a fire engine, which he sent for Franklin's inspection; and "though the original purpose for which the engine had been contrived was the pumping of water, it was believed to be practicable to employ it as a means of locomotion. Franklin was too much occupied by grave political questions to pursue the subject;

but the sanguine and speculative mind of Erasmus Darwin was inflamed by the idea of the 'fiery chariot,' and he pressed his friend Boulton to prosecute the contrivance of the necessary steam machine. Erasmus Darwin was, in many respects, a remarkable man. In his own neighbourhood, he was highly esteemed as a physician, and by the most intelligent readers of his day, he was greatly prized as a poet. The doctor was accustomed to write his poems with a pencil on little scraps of paper while riding about among his patients in his sulky. The vehicle, which was worn and bespatted outside, had room within it for the doctor and his appurtenances only. On one side of him was a pile of books, reaching from the floor to nearly the front window of the carriage, while on the other was a hamper containing fruit, &c., with which the occupant of the carriage regaled himself during his journey. Lashed on to the place usually appropriated to the boot was a large pail for watering the horses, together with a bag of oats and a bundle of hay." Such was the equipage, says Dr. Smiles, of a fashionable country physician of the last century. As he drove through the country in his sulky, his mind teemed with speculations on all subjects: though his speculations were not always sound, they were clever and ingenious, and, at all events, they had the effect of setting other minds thinking, and speculating on science and methods for its advancement. It would appear that the doctor even entertained a theory of managing the winds by a little philosophic artifice. His scheme of a steam locomotive was of a more practical character. This idea, like so many others, first occurred to him in his sulky. "As I was riding home yesterday"— he wrote to his friend Boulton in the year 1765 — " I considered the scheme of the fiery chariot, and the longer I contemplated this favourite idea, the more practicable it appeared to me. I shall lay my thoughts before you, crude and undigested as they appeared to me, and by these hints you may be led into various trains of thinking upon this subject, and by that means (if any hints can assist your genius, which, without hints, is above all others I am acquainted with) be more likely to approve or disapprove. And as I am quite mad of the scheme, I hope you will not shew this paper to anyone. These things are required : (1) a rotary motion ; (2) easily altering its direction to any other direction ; (3) to be accelerated, retarded, destroyed, revived, instantly and easily ; (4) the bulk, the weight, the expense of the machine to be as small as possible in proportion to its weight." In this letter Darwin gives numerous sketches. and suggests that the steam carriage should have three or four wheels, and be driven by an engine having two cylinders open at the top, and the steam condensed in the bottom of the cylinder, on Newcomen's principle, which was the engine of his day, used chiefly for pumping water from mines. The steam was to be admitted into the cylinders by cocks worked by the person in charge of the steering wheel, the injection cock being actuated by the engine itself. " And if this answers in practice as it does in theory," he thought " the machine could not fail of success." Darwin's ideas were very crude, but he was most anxious to put his thoughts into practice, requesting Boulton to become a partner with him in the profit, expense and trouble, as he was determined to build a fiery chariot, and if it answered, get a patent. Boulton does not appear to have taken the scheme up, consequently the doctor allowed the matter to lapse. Dr Smiles from whom we have quoted the above says :—" It is clear that even though Boulton had taken up and prosecuted Darwin's idea, which he did not, it could never have issued in a practicable or economical working locomotive."

MOORE.—Francis Moore, a linen draper, in March, 1769, invented a new machine or engine, made of wood, iron, brass copper, or other metals, and constructed upon peculiar principles, and capable of being wrought or put in motion by fire, water, or air, without being drawn by horses, or any other beast, or cattle; and which machines, or engines, upon repeated trials, he has discovered would be very useful in agriculture, carriage of persons and goods, either in coaches, chariots, chaises, carts, wagons, or other conveyances, and likewise in navigation, by causing ships, boats, barges, and other vessels to move, sail, or proceed, with more swiftness or despatch. From a letter which appeared in the *Leeds Mercury* for 11th April, 1769, referring to this new engine to go without horses, for which Moore had obtained His Majesty's patent, we learn that the ingenious inventor had sold all his own horses, and by his advice many of his friends had done likewise, because the price of that noble and useful animal would be so affected by his invention, that their value would not be one-fourth of what it was then.

Inventors are usually very sanguine, but Mr. Moore was exceptionally hopeful, or he would not have sold his own horses; he displayed his faith in his invention in a practical manner, and he must have possessed considerable influence to have enabled him to persuade his friends to follow his example. The steam carriage is one of the many expedients invented for getting rid of the horse, and Moore was under the impression that the universal adoption of his engines would work a revolution, and enable farmers to dispense with horse power for drawing loads. These results, so long ago anticipated by Moore, have been to some extent realized, and the "iron horse" is doing the heavy haulage on our highways, ploughing land, threshing the grain, &c. Moore made repeated trials with his steam carriage, and took out three distinct patents relating thereto, one in March, the second in June, and the third in July of 1769; but as no drawings accompanied the patent specifications, we are unable to give any details respecting his inventions. Dr. Small, of Birmingham, in writing to Watt, who was then in Glasgow, alluded to Moore's invention as follows : " A linen draper at London, one Moore, has taken out a patent for moving wheel carriages by steam. This comes of thy delays. I daresay he has heard of your inventions." Watt, in reply to Dr. Small, says : " If linen draper Moore does not use my engine to drive his chaises, he cannot drive them by steam. If he does I will stop him. I suppose by the rapidity of his progress and puffing, he is too volatile to be dangerous."

Watt then reverts to the folly of inventing, and imagines Moore hiring 2,000 men, setting them to work on his steam carriages, and making a fortune ; but Moore's fortune was not made by the manufacture of road locomotives. We read that he was rich and sanguine ; and Dr. Small recommends Watt to ask an exorbitant price if Moore writes for permission to use his engine. Watt received no royalty from Moore, and had no occasion to stop his steam carriages, as they soon ceased to attract any further attention.

SMALL.—Dr. Small, after scolding Watt for his negligence, then tries to persuade him to give attention to this subject, and, with this object in view, gives him the result of his own reflection. " I have thought " adds the doctor, on 5th November, 1769, " of a very easy method of constructing your steam wheel (rotary engine), and of a most easy and obvious method of moving carriages by a reciprocating engine providing a tolerably tight piston can be found. The weight of the machine for a post chaise will not be more than 300 lbs., water and all, and it will be contained in a very small compass."

EDGEWORTH.—Although Darwin allowed the subject of a steam chariot to lapse as far as he was concerned, yet we learn that he succeeded in inflaming the mind of his young friend Richard Lovell Edgeworth, and induced him to direct his attention to the introduction of improved means of locomotion by steam power. In August, 1768, Dr. Small informed Watt that Mr. Edgeworth, a gentleman of fortune, young, mechanical, and indefatigable, had taken a resolution of moving land and water carriages by steam, and had made considerable progress for the time he had employed himself in that line. "He knows nothing" added Dr. Small, "of your peculiar improvements, but seems to be in a fair way of knowing whatever can be known on such subjects." Mr. Edgeworth patented in 1770, a portable railway or artificial road, to move along with any carriage to which it is applied. The invention consisted in making "portable railways to wheel carriages, so that several pieces of wood were connected to the carriage, which it moved in regular succession in such a manner that a sufficient length of railway was constantly at rest for the wheels to roll upon, and that when the wheels had nearly approached the extremity of this part of the railway their motion laid down a fresh length of rail in front, the weight of which in its descent assisted in raising such part of the rail as the wheels had already passed over; and thus the pieces of wood which were taken up in the rear were in succession laid in front, so as to furnish constantly a railway for the wheels to roll upon." Mr. Edgeworth has had hundreds of persons who have imitated his schemes, and revived the portable railway again and again. Mr. Boydell was the one who approached the nearest to success; his wheels were largely adopted, and appeared to give fair satisfaction; but they were too complicated and cumbersome, and the wear and tear was something ruinous, causing them to be soon discarded. Notwithstanding the numerous failures, this is still a favourite subject with inventors, and we are at the present day occasionally called upon to examine some of these ridiculous schemes. Steam carriages will run well on proper roads, but they are not intended to travel over swamps, and it is scarcely likely that any application of a patent railway will make them succeed where proper roads do not exist.

Mr. Edgeworth read a paper before the Society of Arts on Railroads, for which he was awarded the society's gold medal. In this paper he proposed four iron railroads to be laid down on one of the great roads out of London—two for carts and heavy vehicles passing in either direction, and the other two lines to be used for lighter traffic, gentlemen's carriages, post chaises, all to be drawn by horses. The post chaises, he thought, might travel at eight miles an hour, and the stage coaches at six miles an hour. In 1802, he suggested the use of Watt's stationary engines, fitted with winding gear for drawing the trains by ropes or chains. Mr. Edgeworth revived the sailing carriage scheme, and, after having patented one, he carried out a number of experiments; but like his predecessors, he soon found the wind to be a very unsuitable power. Sometimes there was no wind, at others it blew from the wrong quarter, and when it so happened that it blew strong in the right direction, his carriages appeared to fly along the line.

In Vol. I. of Edgeworth's Memoirs it is stated: "Even the method of locking carriages, in turning, invented by Dr. Darwin and myself, had been employed in a sailing carriage, described in the 'Machines Approuvées' of the French Royal Academy." Mr. Edgeworth tells us that he spent forty years in experimenting with his portable railway, without being able to gain the necessary strength and lightness desired. He says: "As an encouragement to perseverance, I assure my readers that I have made considerably over one hundred working models upon the portable railway system in a great variety of forms"; and, although he had not been able to accomplish his object, yet he was satisfied that the scheme was feasible. "The experience which I have acquired by this industry has overpaid me for the trifling disappointments I have met with, and I have gained far more in amusement than I have lost by unsuccessful labour. Indeed, the only mortification that affected me was my discovering, many years after I had taken out my patent, that the rudiments of my whole scheme were mentioned in an obscure memoir of the French Academy." Now in Vol. III. of the "Machines Approuvées par l'Academie Royale des Sciences," there is a description of a sailing carriage, which, in the act of turning, rests upon four points of support, like the carriage invented by Dr. Darwin and Mr. Edgeworth; and in the same volume there is an account of a truck, or low carriage, for heavy weights, provided with small rollers, or wheels which travel upon an endless chain of rollers. These are the descriptions referred to. Mr. Edgeworth continued to take pleasure in mechanical matters until he was far advanced in life. He proposed making a cast iron tunnel for crossing the Menai Straits instead of the plan Mr. Rennie had then proposed of a bridge. When an old man of seventy, he wrote to James Watt, 7th August, 1813: "I have always thought that steam would become the universal lord, and that in time we should scorn the post horses." Dr. Smiles says:—" Four years later he died, and left the problem which he had nearly all his life been trying ineffectually to solve, to be worked out by younger men. Dr. Darwin, his co-worker, had long before preceded him into the silent land. Down to his death in 1802, Edgeworth had kept up a continuous correspondence with Watt on his favourite topic."

THE PERIOD OF EXPERIMENT.

CUGNOT.—It is interesting to relate that the first steam locomotive engine which actually carried passengers on common roads, was made by an ingenious French mechanic Nicholas Joseph Cugnot, a native of Void, in Lorraine, where he was born in 1729. In his youth, Cugnot served in Germany as an engineer. He published several works on military science, and served in both the French and German armies. The invention of a light gun procured him the notice of the Comte de Saxe, to whom, Stuart says, in 1763 he exhibited a model of a carriage to be moved by a steam engine, instead of horses. No mention is made, however, of this model of 1763 in General Morin's exhaustive paper in the *Comptes Rendus* for 14th April, 1851. Cugnot afterwards lived at Paris, and through the recommendation of the Comte, obtained the patronage of the Duke de Choiseul, then Minister of War. After his retirement from the army, he was enabled, at the public expense, to construct a steam carriage to run on common roads, which was tried at the Arsenal in 1769, in the presence of the Duke de Choiseul, General Gribeauval First Inspector-General of Artillery; Count Saxe, and other distinguished personages. It was mounted upon three wheels, the leading wheel being driven by an engine whose two pistons acted upon it alternately. During the time that Cugnot was making his steam carriage, a Swiss officer named Planta showed the Duke de Choiseul a model of a similar machine; but, being commissioned by General Gribeauval to examine Cugnot's machine, and having found it in every way superior to his own, he relinquished the idea of making a carriage on his plans. During the first run, Cugnot's machine carried four persons, and travelled at the rate of two and a quarter miles an hour. The boiler, however, being too small, the carriage could only run for fifteen or twenty minutes before the steam was exhausted, and it was necessary to stop the engine for nearly the same time, to enable the boiler to raise the steam to the maximum pressure, before it could proceed on its journey. Thurston says: "This machine was a disappointment, in consequence of the inefficiency of the feed pumps."* Few trial runs have been made with road locomotives without some accident having to be chronicled in the proceedings, and Cugnot's engine was no exception to the rule. Very early in its career, it displayed its power by knocking down a stone wall which happened to be in its way. This casualty naturally caused unfavourable impressions to be formed respecting it; those who witnessed the test said that its use was attended with some uncertainty, and urged that unless its motion were placed under proper control, it could not be of use in practice. Considering the time of its appearance, this first attempt was remarkable, and the trial of the pioneer road locomotive cannot be regarded as an unsuccessful one. No sooner had the machine emerged from the workshop, than the idea occurred to some that, provided it could be made more powerful, and its mechanism improved, it might be used to drag cannon into the field instead of using horses for this purpose. Consequently, Cugnot was ordered by the Minister of War to proceed with the construction of an improved and more powerful

* "History of the Growth of the Steam Engine." (Thurston 1879.)

machine, which was finished and ready for trial during the latter part of the year 1770, having cost 20,000 livres. The actual machine is still preserved in an old church attached to the Conservatoire des Arts et Metiers, in Paris, and is in an excellent state of preservation. The carriage and its machinery are substantially built, well finished, and show exceedingly creditable work in every respect. It is, without exception, the most venerable and interesting of all the machines extant connected with the early history of locomotion. Figs. 2 and 3 show a side elevation and plan, of Cugnot's improved steam road locomotive. It consisted of two parts—one of a carriage on two wheels intended for the load, furnished with a seat for the steersman; the other of the engine and boiler, supported on a single driving wheel 4ft. 2in. diam., this front part taking the place of the horse. The two parts are united by a movable pin. A toothed quadrant, fixed on the framing of the fore part, is actuated by spur gearing on the upright steersman's shaft in close proximity to the seat, by means of which the conductor could cause the carriage to turn in either direction, at an angle of from 15° to 20°. The front part, as will be seen, carries the round copper boiler, having a furnace inside, provided with two small chimneys, the two single acting brass cylinders, 13 in. diam. communicating with the boiler by the steam pipe, fitted with a four-way cock, and the other machinery for communicating the motion of the pistons to the driving wheel. On each side of the driving wheel, ratchet wheels are fixed, and as one of the pistons descends, the piston rod draws with it a crank, the pawl of which, working into the ratchet wheel, causes the driving wheel to make a quarter of a revolution. By means of suitable gearing, the same movement also places the piston

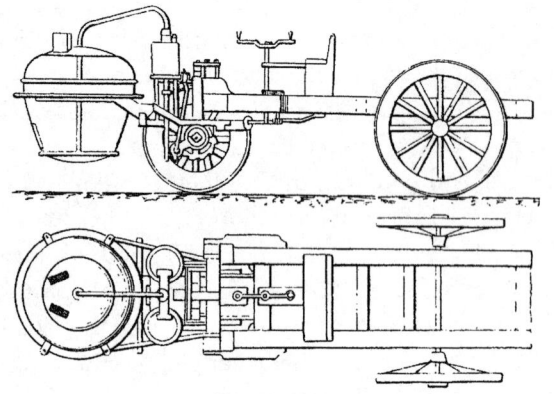

Fig. 2 & 3.

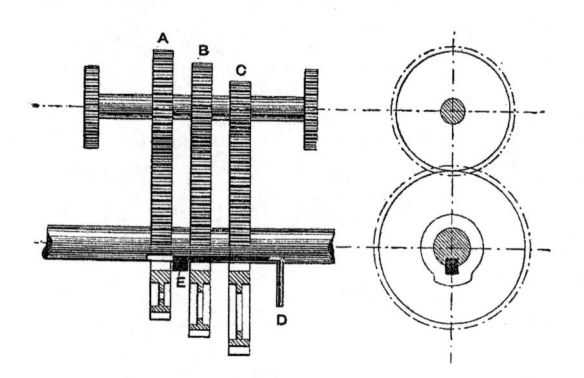

Fig. 5.

on the other side in a position for making a stroke, and turns the four-way cock, so as to open the second cylinder to the steam and the first cylinder to the atmosphere. The second piston then descends, causing the leading wheel to make another quarter of a revolution, and restoring the first piston to its original position. In order to make the vehicle run backwards, the pawl being arranged to act either above or below, it was merely necessary to make it act on the upper side (changing the position of the spring which pressed upon it); then, when the engine was started, the pawl caused the driving wheel to turn a quarter of a revolution in the opposite direction with every stroke of the piston.[*] It will be noticed that Cugnot's carriage was, more than a century ago, actuated by a simple and ingenious form of high pressure engine. Several successful trials were made with the new locomotive in the streets of Paris, exciting no small degree of interest. Unhappily, the machine met with an accident during one of its journeys, which brought the trials to an untimely end. Running upon three wheels only, with the weight of the boiler and engine overhanging in front, it was by no means a steady machine, and in passing along a street in Paris at the rate of three miles an hour, near where the Church of the Madeleine now stands, when turning a corner it overbalanced itself, and fell over with a crash. It was consequently locked up, to keep it out of further mischief, and poor Cugnot was also imprisoned. The merit of Cugnot was, however, duly recognised, and he was granted a small pension, which continued to be paid to him until the outbreak of the Revolution. He then suffered the keenest privations,

[*] A more detailed description of Cugnot's locomotive is given in a paper read by E. A. Cowper, Esq., C.E., before the 'Institution of Mechanical Engineers,' in 1853. Dr. Smiles also gives some particulars of this relic and its inventor in "*The Engineer*," 21st December, 1866.

such as have been the fate of too many inventors, who, like Cugnot, were unfortunately half a century ahead of their contemporaries. Napoleon eventually restored Cugnot's pension, and thus soothed his declining years. He died in Paris on the 10th October, 1804, at the advanced age of seventy-five.

MURDOCK. — Among those who have contributed largely towards making steam locomotion on common roads an accomplished fact, a prominent place is occupied by William Murdock, Messrs. Boulton & Watt's ingenious and esteemed assistant, who was born August, 1754. During his early years he worked for his father, in the mill on the farm, and assisted in the preparation of mill machinery at Bellow Mill, near Old Cumnock, in Ayrshire; between them they constructed a wooden horse, worked by mechanical power, on which young Murdock travelled about for miles around the neighbourhood of his native place, to the amazement of the inhabitants of the district. Having often heard of the inventions of Watt, he determined to seek employment at the famous works at Soho; and, in order to carry out this intention, he left Scotland in 1777, in the twenty-third year of his age. Upon his arrival at the works he gained an interview with Mr. Boulton, who, after asking a few questions, gave him no encouragement, telling him they had no vacancy, as trade was slack, and was bidding him good speed to some other shop. Just as the footsore and shabbily dressed millwright was turning sorrowfully away, Boulton was struck with the peculiar appearance of the hat Murdock wore, and suddenly called him back.

"That seems to be a curious sort of hat," said Boulton, looking at it more closely; "what is it made of"? "Timmer, sir," said Murdock modestly. "Timmer? Do you mean to say it is made of wood?" "Deed it is, sir." "And pray how was it made?" "I just turned it in the lathie." "But its oval man, and the lathe turns things round." "Aweel! I just gar'd the lathie gang anither gate to please me. I'd a long journey afore me, and I thocht to have a hat to keep out water, and I hadna muckle siller to spare, and I made me ane." Boulton looked at the young man again. He had risen a hundred degrees in his estimation. Murdock was a good-looking fellow—tall, strong, and handsome—with an open intelligent countenance. Besides, he had been able to turn a hat for himself with a lathe of his own construction. This, of itself, was a sufficient proof that he was a mechanic of no mean skill. "Well!" said Boulton at last, "I will enquire at the works, and see if there is anything we can set you to. Call again, my man." "Thank you sir," said Murdock. When he called again he was put upon a trial job. Such was the beginning of Murdock's connection with the celebrated firm of Boulton & Watt.

Being found a satisfactory workman he was engaged for two years at fifteen shillings per week. He applied himself diligently and conscientiously to his work, and gradually rose from grade to grade, until he became a most trusted co-worker and adviser in all his employers' mechanical undertakings. Murdock, doubtless, had heard of Watt's proposed methods of applying the steam engine to the propulsion of carriages on common roads; and while Watt was discussing matters of detail respecting his carriage with Boulton, their ingenious workman was busily employing his leisure hours, while residing at Redruth, in 1781, in the construction of a model locomotive to his own ideas, the result being the beautifully simple little high pressure non-condensing engine we are all so familiar with, which was doubtless, the first successful locomotive ever made in England. This engine of liliputian dimensions was ready for trial in 1784. The first experiment was made in Murdock's house at Redruth, when the locomotive successfully hauled a waggon round the room, the single wheel, placed in front of the engine, fixed in such a position as to enable it to run round a circle.

Dr. Smiles says:—" Another experiment was made out of doors, on which occasion, small though the engine was, it fairly outran the speed of its inventor. One night, after returning from his duties at the mine at Redruth, Murdock went with his model locomotive to the avenue leading to the church, about a mile from the town. The walk was narrow, straight, and level. Having lit the lamp, the water soon boiled, and off started the engine with the inventor after it. Shortly after he heard distant shouts of terror. It was too dark to perceive objects, but he found, on following up the machine, that the cries had proceeded from the worthy vicar, who, while going along the walk, had met the hissing and fiery little monster, which he declared he took to be the Evil One in *propria persona!*"*

Fig. 4 represents Murdock's interesting little locomotive, from which it will be seen that it comprises a flat board, at one end of which is a wood upright, carrying one end of the beam. The cylinder is $\frac{3}{4}$in. diam., with a stroke of piston of 2 in., and placed underneath the opposite end of this beam, to which is likewise attached the connecting rod for actuating the crank on the axle of the $9\frac{1}{4}$in. diam. driving wheels. The double cylindrical slide valve is worked by a tappet motion, the beam striking the shoulders of the valve spindle, and forcing it up and down alternately at each end of the stroke, and the steam is exhausted through the hollow spindle of the valve, passing out near the top. The disc seen round the

* Invention and Industry.—Dr. Smiles, 1884.

MURDOCK'S LOCOMOTIVE, 1781.

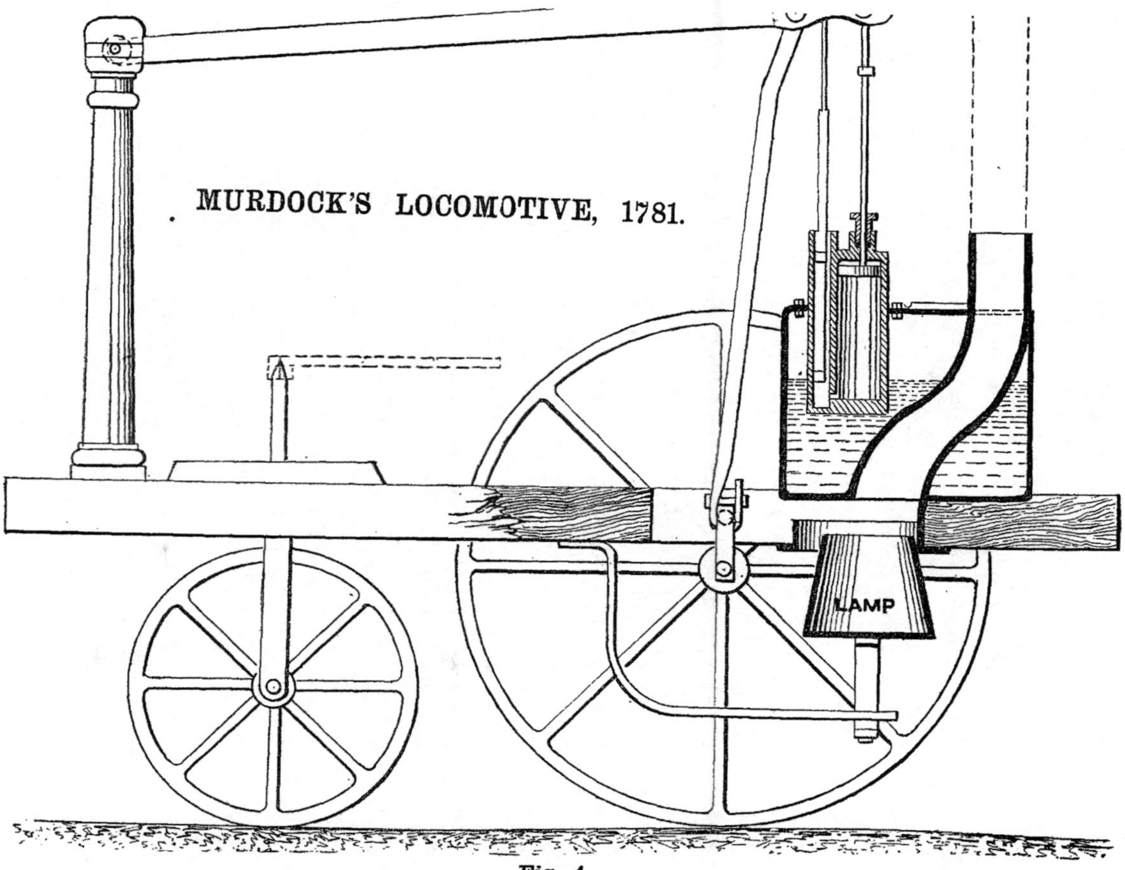

Fig. 4.

vertical pivot of the steering wheel is a lead weight, apparently put on to keep the front of the engine down, and make it answer more easily to the steerage wheel, which works in a swivel frame, to allow the engine to turn in a small circle. The boiler is made of copper, with the flue passing obliquely through it, as shown in section on the drawing, and heated by a spirit lamp. The fact of one of the driving wheels only being fixed on the axle accounts for the carriage travelling in zigzags. This interesting little relic was made by Murdock's own hands, and repeatedly shown by him to friends at his house at Handsworth, up to the time of his death.

Mr. Richard Tangye in "One and All" says:—"The model has been continuously in the possession of the Murdock family till 1883, when it was purchased from Murdock's great grandson by Messrs. Richard and George Tangye, and lent by them to the Melbourne Exhibition in 1889, where it was exhibited alongside Symington's model. It is now in the Birmingham Art Gallery."

It is still in good working order, although over one hundred years old, and when under steam is capable of attaining a speed of 6 to 8 miles an hour. Watt was very displeased when he heard of Murdock's experiments with the road locomotive, because he feared that it might have the effect of withdrawing his mind from his business, and causing him to give less attention to the interests of the firm ; for, at that time, Murdock was an indispensable member of their staff, capable of undertaking the most difficult task connected with the mining engines and machinery erected by Boulton and Watt; he was also a general favourite among the miners. Watt accordingly wrote to Boulton, recommending him to advise Murdock to give up his locomotive engine scheme ; but if he could not succeed in that, then, rather than lose Murdock's services, Watt proposed that

THE PERIOD OF EXPERIMENT.

"he should be allowed an advance of £100 to enable him to prosecute his experiments, and if he succeeded within a year in making an engine capable of drawing a post chaise, carrying two passengers and the driver at four miles an hour, it was suggested that he should be taken as partner into the locomotive business, for which Boulton and Watt were to provide the necessary capital."* This proposition was never carried out. Again, in 1786. we find Watt expressing his regret that Murdock was "busying himself with the steam carriage," and further on in the letter he says, "I wish William could be brought to do as we do, to mind the business in hand, and let such as Symington and Sadler throw away their time and money in hunting shadows." Murdock applied himself to the business, but continued to speculate about steam locomotion on common roads, he being persuaded of its practicability, but refrained from embodying his ideas in a more complete form.

"In 1815, while Murdock was engaged in erecting an apparatus of his own invention for heating the water for the baths at Leamington, a ponderous cast-iron plate fell upon his leg above his ankle, and severely injured him. He remained a long time at Leamington, and when it was thought safe to remove him, the Birmingham Canal Company kindly placed their excursion boat at his disposal, and he was conveyed safely homeward. So soon as he was able, he was at work again in the Soho factory. In the midst of repeated inventions and experiments, Murdock was becoming an old man. Yet he never ceased to take an interest in the works at Soho. In his latter years his faculties experienced a gradual decay, and he died peacefully at his house at Sycamore Hill, on the 15th of November, 1839, in his eighty-fifth year, and his remains were accompanied by several old and attached friends and the Soho workmen, to their last abode in

* Invention and Industry.—Dr. Smiles 1884.

Handsworth Church, and are there laid near those of Mr. Boulton and Mr. Watt."*

WATT.—In dealing with the early days of steam transport on common roads, we are anxious to award the amount of credit justly due to each promoter in proportion to the services rendered — according honour to whom honour is due. It is very well known that Watt all through his life opposed rather than aided steam locomotion on common roads, and yet, by some authors, if Watt is not credited with the invention of the steam carriage, he is looked upon as the inventor of the practical means for effecting locomotion on land by steam. He was repeatedly urged by his friends to turn his attention to the subject, but without effect. Dr. Small and other correspondents were continually sending him their ideas upon the subject, with a view to coax him to work out the problem. Watt was displeased with Murdock for spending his leisure hours in the construction of a successful locomotive; he threatened to stop Moore from making steam carriages; and, in 1786, he caused Sadler to discontinue any further experiments in the application of the steam engine to the driving of wheel carriages. Watt displayed a dog-in-the-manger spirit; he could not or would not make a steam carriage himself, so appeared determined to let no one else succeed; he even confessed that the specification of his steam carriage was very defective, but it would only "serve to keep other people from similar patents." Dr. Smiles, in his "Men of Invention and Industry," says that Watt, " when only twenty-three years of age, at the instigation of his friend Robinson, made a model locomotive, provided with two cylinders of tin plate; but the project was laid aside and never again taken up by the

*Dr. Smiles, also Mr. Buckle's paper read before the "Institute of Mechanical Engineers," entitled, "The Inventions and Life of William Murdock."

inventor." This model locomotive is not mentioned by Muirhead in his "Life of Watt," and we have reasons for questioning the accuracy of the statement. We believe Watt's plans were never sufficiently matured to enable him to carry them into effect, and, judging from his writings, we imagine he was, at the end of his career, no nearer the construction of a steam carriage, than when he embodied his ideas on the subject in his patent specification. We shall now briefly refer to Watt's ideas on steam locomotion, as contained in the well known specification, No. 1432, dated August, 1784. He says: "My seventh new improvement is upon steam engines which are applied to give motion to wheel carriages, for removing persons or goods from place to place." For the sake of lightness the boiler was to be made of wood, or of thin metal sheets strongly hooped, of cylindrical or globular form. The fire was contained in a vessel within the boiler. He proposed the use of one or two cylinders, the pistons of which were moved by the steam in their ascent and descent. The exhaust was to be either discharged into the atmosphere by a regulating valve, or else conducted into an air condenser, formed of thin plates, or pipes of metal having their outsides exposed to the wind, or cooled by a bellows or fan driven by the engine; this condenser adding to the power of the engine and economising the condensed steam, which would otherwise be lost. Motion was to be communicated from the pistons to the intermediate shaft by means of his sun and planet wheels; and Watt very sensibly proposes the use of spur gearing, for communicating the power from the intermediate shaft to the main axle, by means of which he readily introduces pinions and wheels of various sizes for obtaining different travelling speeds; any pair of these wheels being put into gear by a clutch of simple construction, to which we shall refer shortly. The carriage, intended for two persons, would require an engine having a cylinder 7in. in diam., with a stroke of 12in., making sixty strokes per minute, the steam pressure to be "equal to supporting a pillar of mercury 30in. high." Although Watt promised to construct a model locomotive in accordance with the specification, yet he never fulfilled the promise, his time apparently being more than fully occupied with other work; moreover, his mind was unable to embrace the idea that steam locomotion would ever be accomplished. A few days after filing his specification, Watt wrote to Boulton, and entered more minutely into the details of his proposed steam carriage, wherein he gives a sketch of the spur gearing we have favourably mentioned above. Fig. 5 represents this gearing, from which it will be seen that three different sized wheels are keyed fast on the shaft, which receives the motion from the sun and planet gearing. These wheels are made to gear with others running loosely on the main axle, any pair of which can be made to drive the carriage faster or slower (according to the nature of the roads to be traversed, or the loads to be drawn), by means of a clutch or feather sliding along the axle, and made to fit into the bosses of the wheels. When the clutch is made to engage with one pair of these wheels, motion is imparted through the train of gearing from the intermediate shaft to the main axle. It will be seen that the sliding key or clutch must be disengaged from one wheel before it is made to interlock with another. This idea of Watt's is a decided advance upon all previous methods of communicating motion from the crank shaft to the axle, it being the plan adopted on all the road locomotives of the present day. In the same letter he proposes to make the carriage run at four miles an hour; the driving wheels are to be 4ft. diam., and the crank shaft is to make twenty revolutions to one of the main axle. The proper place for the engine, Watt thought, was behind the carriage. Another plan, something after Cugnot's, was proposed, of having two inverted cylinders with toothed racks instead of piston rods, which were to be applied to two ratchet wheels on the main axle, and to act alternately. "I am partly of opinion," says Watt, "that this method may be applied to advantage yet, because it needs no fly, and has some other conveniences." In October, 1786, Watt writes to Dr. Black, and intimates that he will carry out some experiments with wheel carriages moved by his steam engine; but he entertains small hopes of their ever becoming useful. During the same year, in writing to Boulton, it would appear as though Watt had a steam carriage in hand; but soon afterwards the carriage schemes were allowed to lapse altogether. It will thus be seen that Watt's ideas on this subject were very crude, and had he attempted to put his ideas into practice, we fear the machine would have been a failure, with its low pressure wooden boiler, air condenser, and sun and planet wheels;* but it is well known that Watt retained, up to the period of his death, the most rooted prejudices against the use of high pressure steam. He says; "I soon relinquished the idea of constructing an engine on this principle, from being sensible it would be liable to some of the objections against Savery's engine, viz., the danger of bursting the boiler," and also that power would be lost without the use of a vacuum. He, therefore, can only be said to have proposed steam locomotion on common roads, and it is undoubtedly true that Watt never built a steam carriage.

SYMINGTON.—William Symington, the engineer, who is generally acknowledged to be the inventor of the first success-

* "We have seen that great, if not insurmountable difficulties are overcome by the simple apparatus of a high pressure engine; but how a condensing engine would have succeeded, when encumbered with its load of condensing apparatus, requires no very extraordinary discernment to discover."—"History and Progress of the Steam Engine," by Galloway.

ful steam boat, was born at Leadhills, in Scotland, in 1763 His father was a practical mechanic, who superintended the engines and machinery of the Lead Mining Company at Warlockhead, where one of Boulton and Watts' pumping engines was at work. Young Symington, like Murdock, was placed under his father's tuition in the north country workshop, and, like Murdock also, he gave early proof of his ingenuity. He appears at the age of twenty-one to have conceived the idea of applying the steam engine to the propulsion of carriages; his father and he worked together to carry the idea into effect, and by the year 1786 they succeeded in completing a working model of a road locomotive. So efficiently did the model work, that those who saw the machine expressed such favourable opinions respecting it that the difficult problem of moving carriages on the highway by steam power appeared to be within measurable distance of being solved, and Symington was warmly urged to carry his experiments to a practical issue. Mr. Meason, the manager of the lead mine, "was so pleased with the model, the merit of which principally belonged to young Symington, that he sent him into Edinburgh, for the purpose of exhibiting it before the professors of the University, and other scientific gentlemen of the city, in the hope that it might lead in some way to his future advancement in life." Moreover, Mr. Meason, who proved to be his patron and friend, allowed the model to be exhibited at his own house, and invited many persons of distinction to inspect it, and he liberally offered to defray any expense which might be incurred in carrying the invention out in practice. The state of the roads, and the difficulty which at that time existed of procuring water and fuel, afforded sufficient reasons to induce Symington to conscientiously abandon the scheme, which through these causes, he believed, would only have produced disappointment to his kind advisers. By referring to the illustration of Symington's locomotive, Fig. 6, it will be seen that it consisted of a carriage with a locomotive behind, supported on four wheels, the front wheels being arranged with steering gear. A cylindrical boiler was used for generating steam, which communicated by a steam pipe with the two horizontal cylinders, one on each side of the firebox of the boiler. When steam was turned into the cylinder, the piston made an outward stroke; a vacuum was then formed, the steam being condensed in a cold water tank placed beneath the cylinders, and the piston was forced back by the pressure of the atmosphere. The piston rods communicated their motion to the driving axle and wheels through rack rods, which worked toothed wheels placed on the hind axle on both sides of the engine, and the alternate action of the rack rods upon the tooth and ratchet wheels with which the drums were provided, produced the rotary motion. Symington stated that a material advantage obtained by the mode here employed of applying the power of the engine, was that it always acted at right angles to the axle of the carriage. The boiler it will be seen was fitted with a lever and weight safety valve, and as a whole, the arrangement of the engine and carriage displays much ingenuity; but we fear that the rack rods would have proved unsatisfactory, while the travelling speed must have been very slow indeed. Symington's road locomotive was allowed to slumber, never to have an awakening, while the inventor turned his attention to the propulsion of vessels by steam. Mr. Miller, of Dalswinton, in 1787, was making experiments with pleasure boats propelled by paddles worked by manual power, and he complained to Symington of the fatigue caused to the men by working the capstan. Symington at once suggested the adoption of the steam engine for the purpose, and was commissioned by Miller to construct a small steam engine for

SYMINGTON'S MODEL STEAM COACH, 1786.

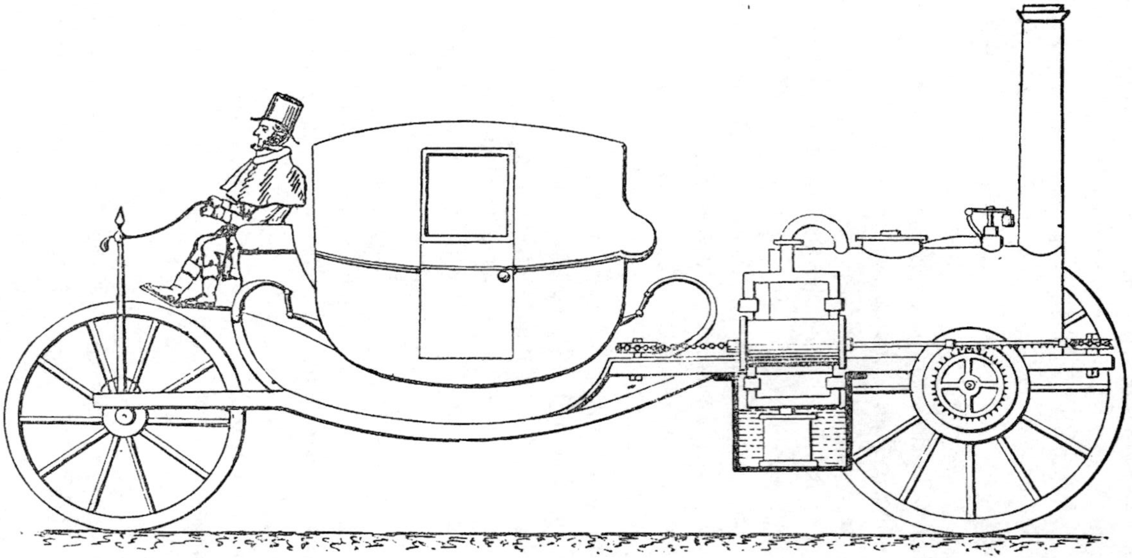

Fig 6.

this pleasure boat, which was finished during the latter part of the year 1788. Many experimental trips were made with this boat, which steamed at the rate of five miles an hour to the delight of Miller and his numerous friends. On the approach of winter the engine was removed from the boat and placed in the library at Dalswinton, and is now in the South Kensington Museum, it having been aptly termed "the parent engine of steam navigation." The year afterwards a larger boat was fitted with an engine made by Symington with a cylinder 18 inches diameter, which on trial, after sundry alterations had been effected, made seven miles an hour. Miller abandoned steam navigation after expending £30,000 in experiments. In 1801, Symington, patronised by Lord Dundas of Kerse, started a steam boat on the Forth and Clyde Canal for the purpose of towing boats which was a complete success, and is now regarded as the *first practical steam boat ever built.* The engine employed in the 'Charlotte Dundas' was made in accordance with Symington's patent taken out in 1801, described as a steam boat engine without a working beam. A Watt's double acting horizontal cylinder being connected to a crank on the paddle shaft at the stern of the vessel.

The Duke of Bridgewater gave Symington an order for eight boats like the 'Charlotte Dundas' to be used on his canal. The death of the Duke, however, prevented the contract from being carried into effect. Poor Symington now shared the fate of many other inventors, he had no capital of his own to work upon, and he could not persuade any capitalist to come to his aid, and "fortune appeared to run steadily against him." Dr. Smiles says:—" The rest of his life was for the most part thrown away. Towards the close of it his principal haunt was London, amidst whose vast population he was one of the many waifs and strays. He succeeded in obtaining a grant of £100 from the Privy Purse in 1824, and afterwards an annuity of £50, but he did not live long to enjoy it, for he died in March, 1831, and was buried in the churchyard of St. Botolph, Aldgate, where there is not even a stone to mark the grave of the inventor of the first practicable steam boat."* Since the above was written, an effort has been set on foot for erecting a suitable memorial to William Symington, which will doubtless be carried out. We allow our benefactors while living to suffer the keenest poverty, and to atone for such cruel conduct we rear statues to their memory when dead.

SADLER.—James Sadler, of Oxford (whose name Watt coupled with Symington's as that of "hunters of shadows"), invented a rotary engine, and afterwards patented a double-cylinder engine, a combination of Cartwright's and Watt's engines. In 1786, Sadler was making some experiments in applying the steam engine to the driving of wheel carriages. Upon Watt hearing of this, notice was forthwith given to him to the effect that "the sole privilege of making steam engines by the elastic force of steam acting on a piston with or without condensation, had been granted to Mr. Watt by Act of Parliament, and also that, amongst other improvements and applications of his principle, he had particularly specified the application of steam engines for driving wheel carriages, in a patent which was taken out in the year 1784." As we can find no record of Sadler's steam carriage, it is most likely that Watt's caution had the effect of putting a stop to further experiments in this direction. Watt was unable to build a steam carriage himself, but he was very anxious that no one else should. Sadler's rotary engine is illustrated and described in "Stuart's History of the Steam Engine," 1824.

* *The Engineer*, 21st December, 1866.

EVANS. — Oliver Evans, a mechanic, who has been designated the Watt of America, was born at Newport, Del., in 1755. He was apprenticed to a wheelwright, and soon exhibited considerable ingenuity and perseverance, coupled with a strong desire to acquire knowledge. In 1772, he endeavoured to discover some means of propelling land carriages without employing animal power; many of the methods before his time, and subsequently tried, presented themselves to his mind, such as the wind, treadles with ratchet wheels, cranks, &c., worked by men; but none of these appliances appeared worthy of practice. His attention was drawn, however, to the possible use of steam for this purpose by the following incident:—"One of my brothers," he says, "had one day been in company with a neighbouring blacksmith's boy, who, for amusement, had stopped the touch hole of a gun barrel, then put in a small quantity of water, and ramming down a tight wadding, after which he put the breech end into the smithy fire, when it discharged itself with as loud a report as if it had been loaded with gunpowder." Evans tried at once to harness this (to him, newly discovered power) to the propulsion of his carriages, but without any sign of success, until a book describing the old atmospheric engine came into his possession, and to his astonishment he found that the steam was only used to form a vacuum. Following up the subject with increased ardour, he soon invented his well-known non-condensing engine, in which the power was derived exclusively from high-pressure steam, and proposed the use of this engine for the propulsion of waggons. In 1786, Evans petitioned the Legislature of Pennsylvania for the exclusive right to use his improvements in mill machinery. and to make and use his proposed steam waggons in that State. The Committee heard him patiently while he described his inventions in flour milling; but when he referred to steam waggons, they believed him to be somewhat deranged, because he spoke of machines which they thought were impracticable. The Board granted the privileges he prayed for, respecting the improvements in flour milling machinery, but they quietly ignored his steam carriage schemes altogether. A similar petition was presented to the Legislature of Maryland, and Evans fortunately had a friend on the Committee, who ably pleaded his cause, by stating that no one in the world had thought of moving carriages by steam, and by granting the request, no one could be injured, and there was a prospect of something useful being produced. He therefore obtained the exclusive right to make and use his steam carriages for fourteen years in that State, commencing from May, 1787.

But after scaling one difficulty, Evans was soon overtaken with others; the right of manufacture had been granted him, but no one believed in his carriage schemes, which were condemned by all those from whom he had expected to receive encouragement. For four years Evans travelled from place to place, showing his drawings to manufacturers, and capitalists; but he failed to find a single person willing to speculate in his carriage experiments, or assist him in putting his ideas into practice. In the year 1801, Evans says:—"Never having found a person willing to contribute to the expense, or even to encourage me to risk it myself, it occurred to me that I had not yet discharged my debt of honour to the State of Maryland, by producing the steam waggons; I determined, therefore, the next day to set to work and construct one." He hired workmen, and had made considerable progress in the steam carriage to be driven by his non-condensing engine, when he suddenly changed his plans, thinking that it would be more profitable to adapt his engine to the driving of mills,

He built an engine having a cylinder 6in. diam., and 18in stroke of piston, which he applied to drive a plaster mill with success. Although he allowed the construction of mill engines to supplant his steam waggons, yet he predicted their eventual success, by stating that his engines would propel waggons on common roads with profit; and that the time would come when people would travel in stages moved by steam engines from one city to another, almost as fast as birds can fly. Evans' engines were now successfully employed for a variety of uses, yet men refused to believe in his steam carriages, and he, being very anxious to shew the practicability of his notions, determined to mount upon temporary wheels and axles a large flat bottomed craft he had on order for the Board of Health of Philadelphia for dredging purposes; he then placed one of his engines in the boat, and applied the power of the engine to the wheels, and slowly propelled this odd looking machine, which he named the "Oructor Amphibolis," from his works, over a mile and a half of rough roads, to the river side, where the wheels were removed, and the boat was launched. The same engine was then applied to the paddle wheel at the stern, and drove the craft by water to its destination. The first carriage driven by steam in America was therefore, the curious looking "Oructor Amphibolus" of 1805, which, as its name indicates, was fit for land or water. Evans wrote a standard treatise on the steam engine, and another on millwrighting, and he continued to build steam engines to the end of his life—19th April, 1819. A long description of Evans' undertakings, with an illustration of his non-condensing engine, is given in "Galloway's History of the Steam Engine," and Thurston gives an illustration of the dredging barge being transported on its temporary wheels driven by the steam engine, in his "History of the Growth of the Steam Engine."

FOURNESS.—In 1788 R. Fourness and J. Ashworth took out a patent for an engine specially designed for driving " travelling carriages of every denomination, or giving circular motion without the assistance of a fly wheel, and without being burdened with cold water for condensing the steam to produce a vacuum in the steam cylinders, or having the machine complicated by the apparatus necessary" to the management and regulation of the condensing water. The drawing in the specification of the patent shows the engine and boiler mounted upon the hinder part of a frame, supported by two pairs of road wheels. Motion is communicated from the pistons of the three inverted open ended cylinders to the crank shaft; one or more spur wheels are keyed upon the crank shaft, gearing into others on the driving axle, for giving the various speeds on the road. Steam is admitted to the cylinders through a " rotative cock," into the barrel of which is introduced three branches for the three cylinders. The cock is worked by a wheel upon it, which receives motion from the engine. A similar arrangement is adopted for dealing with the exhaust, which escapes into a tank for heating the feed water, previous to its being pumped into the boiler. The boiler is provided with a safety valve. Only one of the hind wheels is keyed to the main axle for driving, to allow the carriage to turn corners easily. We are not aware that the inventors ever carried their ideas into effect. The appearance of the carriage is somewhat similar to that of Cugnot's.

ALLEN.—The Earl of Caithness, in 1860, read a paper on steam carriages for common roads, before the British Association for the Advancement of Science, in which he mentions a steam carriage invented by Thomas Allen, of London, in 1789, to carry goods and passengers. No mention appears to be made of this inventor by any other writer, and no account can be found of his schemes having ever been matured.

READ.—Nathan Read, the inventor of the vertical multitubular fire box boiler now in general use, was born at Warren, Massachusetts, in 1759. He was a student of medicine, and afterwards a manufacturer of chain cables and other iron work for ships. In the year 1788, he invented the well-known vertical boiler, which is, as he intended it should be, strong, light, compact, as well as safe, suitable for use in steam boats and for steam carriages.* He spent considerable time in constructing and experimenting with steam boats ; but he does not appear to have met with any success. While busily engaged in his attempts to introduce steam navigation, he designed, and in 1790 obtained a patent for, a steam carriage as illustrated by Fig. 7. The carriage is mounted upon four

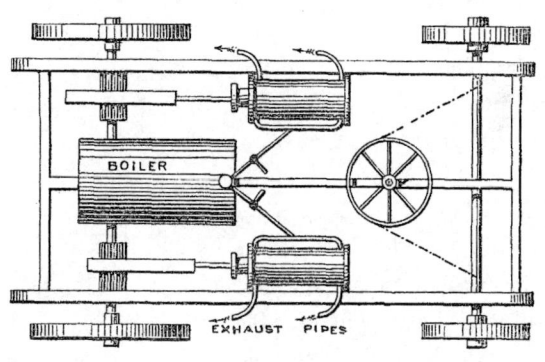

Fig. 7.

travelling wheels ; the hind axle has two ratchet pinions keyed upon it, which are caused to revolve by a rack arrangement attached to each piston rod, by means of which the

* Thurston, in his "History of the Growth of the Steam Engine," gives sectional views of this boiler at page 245. Besides using a series of vertical water tubes, extending from the fire box crown to the water bottom, Read also employed hanging end closed, or "Field" tubes.

carriage is propelled. The boiler is placed in the centre of the carriage over the driving axle, and supplies the two horizontal cylinders with steam, two cocks being inserted in the steam pipes for shutting off the steam from either or both the cylinders. The steerage arrangement is shown, two chains being secured to the front axle, and actuated by the vertical spindle, hand wheel, &c., in the usual manner. The most interesting feature about the locomotive is the arrangement of the exhaust pipes, which the inventor specified should point backwards, so that the reaction of the exhausting steam issuing from the four bent nozzles, as shown should help the carriage forward. Read made a model road locomotive in accordance with his patented plan, which was exhibited during the time that he was seeking assistance in the furtherance of his schemes in road locomotion and navigation by steam power. It is probable, however, that he did not pursue the subject any further than merely making a model. Nathan Read was for some time member of Congress, and during the latter part of his life he filled several influential offices. He died at Belfast, Maine, in 1849, at the age of 90 years.

TREVITHICK.—It is no wonder that the progress made in steam locomotion has been so slow, when we remember that the individuals who have mainly sought to further the subject have possessed very little mechanical knowledge; physicians, for example, a linen draper, a groom, a missionary, wheelwrights, and a chain maker, have each in turn tried their hands at steam carriage construction, and, as may be expected, with anything but satisfactory results. But, when engineers like Cugnot and Murdock handled the subject, progress was at once apparent. Trevithick, however, was the mechanic who did more to bring steam locomotion on common roads to a successful issue than all his predecessors, and if he had received pecuniary assistance, greater success would doubtless have been achieved. As it was, he carried into practice the proposals of Robinson, and fulfilled the predictions of Darwin, Edgeworth, and Oliver Evans. Trevithick was one of the greatest inventors who ever lived; he has been termed the "father of the high pressure engine." In his earliest engines he embodied numerous details which have been in constant use to this day; for example, he heated the feed water by passing the exhaust steam, on its way to the chimney, through tubes round which the feed water circulated, and he turned the exhaust into the chimney to increase the draught. Trevithick made the first road locomotive which conveyed passengers on an English highway. Cugnot had, twenty years previously, carried passengers on his carriage through the streets of Paris; but there was no comparison between these two locomotives. Trevithick's designs displayed a marked advance on all previous types, and his engines were eminently successful. The only reason that we can assign for their non-adoption was the inventor's lack of funds; he could not afford to continue running the carriages at his own expense. Richard Trevithick was born in the parish of Illogan, Cornwall, April 13th, 1771. Shortly after his birth, his family removed to Camborne. We cannot enter into any particulars respecting his early years while at school, or while employed at Stray Park Mine; but we take this opportunity of correcting a misstatement made by Dr. Smiles in one of his excellent books*, which has been copied by several modern writers on the steam engine.†

---

* Life of George Stephenson. † Thurston, in his "History of the Growth of the Steam Engine," says: "Trevithick was naturally a skilful mechanic, and was placed by his father with Watt's assistant, Murdock, who was superintending the erection of pumping engines in Cornwall."

F

Smiles says: "Trevithick was a pupil of Murdock's, and learned from him the knowledge of the steam carriage," which is not correct, as Trevithick was only thirteen years of age when Murdock made his locomotive, and very probably the little model was never seen by Trevithick. As far back as 1796, Trevithick made models of steam locomotives, which were frequently exhibited to his friends at his house at Camborne, and made to run round the table. The boiler and the engine were in one piece; hot water was poured into the boiler, and a red hot iron was inserted into a tube beneath, which caused steam to be raised, and the engine set in motion. In another model the boiler was heated by a spirit lamp. But the most interesting model is the one now in the Kensington Museum, probably made in 1798, and illustrated in Figs. 8 and 9. "It is a perfect

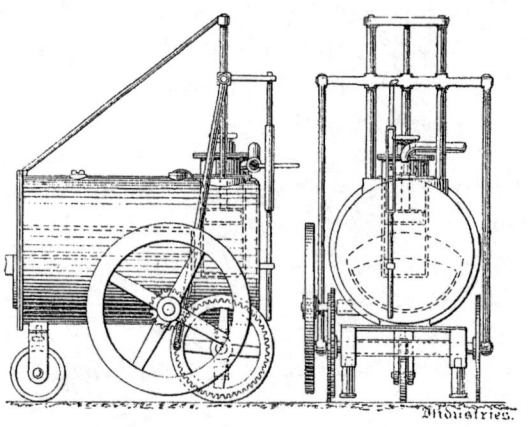

Figs. 8 & 9.

specimen of a high pressure engine, with cylindrical boiler, adapted to locomotive purposes," and served as a guide to those who manufactured engines for Trevithick in later years. The boiler is made of copper, cylindrical in form, with a flue in which was placed a red hot iron for raising steam, so as to avoid smoke. The vertical cylinder is let into the boiler, and with it is cast the shell of the four-way cock, and the pipe for conducting the steam to the top and bottom of the cylinder. The four-way steam cock was worked by a rod from the cross-head. The two side connecting rods are attached to this cross-head at the top, and actuate the two crank pins fixed in the driving wheels. One of these driving wheels has a spur wheel cast with it, which drives a pinion on the fly wheel shaft, as shown. The front wheel works in a swivel frame, and can be fixed in any position for causing the engine to run in a circle. Two legs are attached to the under side of the boiler, which can be screwed out so as to raise the wheels off the ground, and the model then works as a stationary engine. The safety valve is kept down by a simple spring. "The illustrations and description of this working model shadow forth the usefulness of the high pressure steam engine of the present day in many of its leading features. The non-necessity for condensing water, the cylindrical boiler, the simple form of crank, the absence of mason work for the engine or boiler flues, and its portability and power of locomotion, so nearly met all the requirements as to entitle it to the designation, 'the first high pressure locomotive.'" ‡ Trevithick commenced to make his first road locomotive in Tyack's workshop, at Camborne, in November, 1800. An old account book gives some interesting particulars relating to its construction. Several workmen were employed by Trevithick in repairing and improving his mining engines and pumping machinery, and they apparently filled up their time in fitting up and putting the little locomotive together, the workshop being provided with two smiths' hearths and a small hand lathe. The castings were

‡ Life of R. Trevithick. 1872.

made at Harvey's foundry, Hayle, and considerable difficulty was experienced in getting the parts to fit together. Some of the larger turning work was done at Captain Andrew Vivian's workshop. The boiler plates, the steam pressure gauge, and other details came from Coalbrookdale. This pioneer steam road locomotive was completed and ready for testing on Christmas Eve of 1801, and was brought out of the smith's shop on to the highway close by. Considerable interest had been taken by some of the inhabitants during its construction, and, very naturally, these persons were anxious to witness the trial. The following account of the first trial was made by one of the onlookers, who was of the same age as Trevithick, and who resided at Camborne until 1858: " I knew Captain Dick Trevithick very well. I was a cooper by trade, and when Trevithick was making his steam carriage I used to go every day into John Tyack's shop at the Weith, close by here, where they put her together. In the year 1801, upon Christmas Eve, towards night, Trevithick got up steam, out on the high road, just outside the shop. When we saw that Trevithick was going to turn on steam, we jumped up as many as could, may be seven or eight of us. 'Twas a stiffish hill going up to Camborne Beacon, but she went off like a little bird. When she had gone about a quarter of a mile, there was a rough piece of road, covered with loose stones. She didn't go quite so fast, and as it was a flood of rain, and we were very much squeezed together, I jumped off. She was going faster than I could walk, and went up the hill about half a mile further, when they turned her, and came back again to the shop." The next day the engine steamed to Captain Vivian's house for him to see it. A few days subsequently, Trevithick and Vivian started off for Tehidy House, where Lord Dedunstan-ville lived, some two or three miles from Camborne. The former was driving and stoking the engine, the latter was

THE PERIOD OF EXPERIMENT.   53

steering. Just as they were turning a curve in the road, and passing through an open water course across the road, the steering handle was jerked out of Vivian's grasp, and the engine turned over, but in spite of this slight casualty they eventually reached the end of their journey. Several other

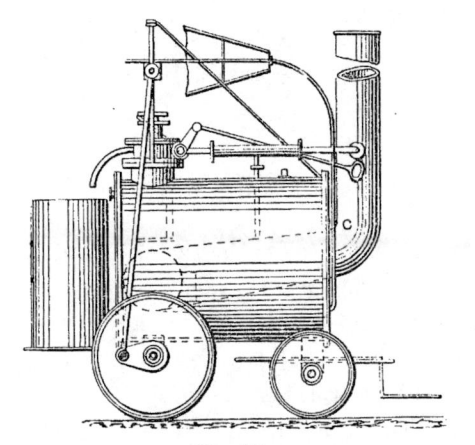

Fig. 10.

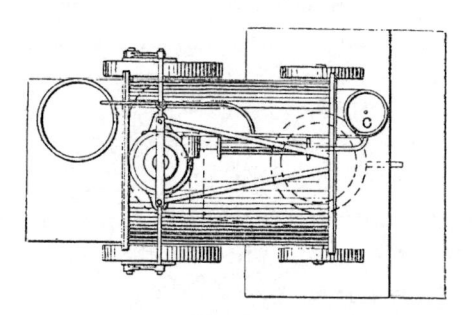

Fig. 11.

trials were carried out, which we need not describe. Trevithick's first passenger steam road locomotive is illustrated in Figs. 10 and 11. The locomotive was extremely simple compared with the engines of that time. Watt's improvements of the last fifty years were discarded, viz., the air pump, condenser, parallel motion, beam, sun and planet wheels, &c. The boiler was cylindrical in form, the shell was made of cast iron, intended for a working pressure of 60lb. of steam; the return flue was made of wrought iron. Fig. 12 shows a

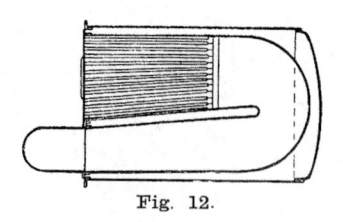

Fig. 12.

section of the boiler. The vertical cylinder was let into the boiler. The exhaust steam, after having done its work in the cylinder, was conducted into the chimney to create a blast, causing an intensely hot fire, and in its passage it heated the feed water. The boiler was provided with a safety valve; a fusible plug was fitted in the crown of the flue, and the feed pump was worked from the cross head. The bellows shown in the side view were not used after the first trial. The four-way cock for admitting steam to the top and bottom of the cylinder was actuated by tappets connected with the cross-head; a lever and handle were also provided, so that the stoker could start or stop the engine. The side rods worked on to cranks keyed on to the driving axle; a lever for steering was conveniently placed near the stoking end of the boiler. The following account respecting the trial of the Camborne steam carriage appeared in a Falmouth paper: "In addition to the many attempts that have been made to construct carriages to run without horses, a method has been lately tried at Camborne that seems to promise success. A carriage has been constructed containing a small steam engine, the force of which was found sufficient, upon trial, to impel the carriage, containing several persons, amounting to a total weight of 30 cwt., against a hill of considerable steepness, at the rate of four miles an hour; and upon a level road of eight or nine miles an hour." The only drawback appeared to be that the boiler was too small, and despite the use of the steam blast in the chimney, steam could not be kept up for long when the carriage was under way. Trevithick and Vivian then became partners, during the last days of 1801, and started off to London to secure a patent for high pressure steam engines for propelling carriages on common roads, &c., which was granted on the 24th March, 1802. The Camborne locomotive was much superior to anything that had previously been attempted; but Trevithick's London steam carriage of 1803, being carried out in accordance with the patent specification, showed a still greater improvement. " It was not so heavy; and the horizontal cylinder, instead of the vertical, added very much to its steadiness of motion; while wheels of a larger diameter enabled it the more easily to pass over rough roads which had brought the Camborne one to a standstill." The boiler was made entirely of wrought iron, and the engine was attached to it, as shown clearly in the illustrations Figs. 13 and 14. The cylinder was inserted in the boiler horizontally, close behind the driving axle. A forked piston rod was used, which, in those days, was a mechanical novelty, the ends working in guides, so that the crank axle might be brought near to the cylinder. Spur gearing and couplings were used on each side of the carriage for communicating motion from the crank shaft to the main driving axle. The driving wheels were about 10ft. diam., and made of wood. The framing was of wrought iron. The coach was intended to seat eight or ten persons, and the greater part of the weight came on the driving axle. The coach was suspended upon springs. The London steam carriage was put together at Felton's carriage shop in Leather Lane, and after its completion, Vivian one day ran the locomotive "from Leather Lane, Gray's Inn Lane, on to Lords' Cricket Ground, to Paddington, and home again by way of Islington, a journey of ten miles through the streets of London." Several trips were made in Tottenham Court

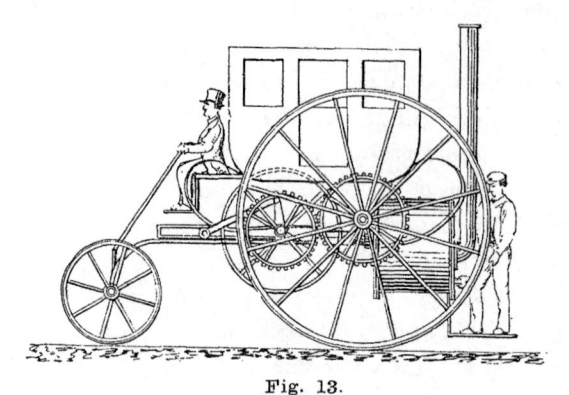

Fig. 13.

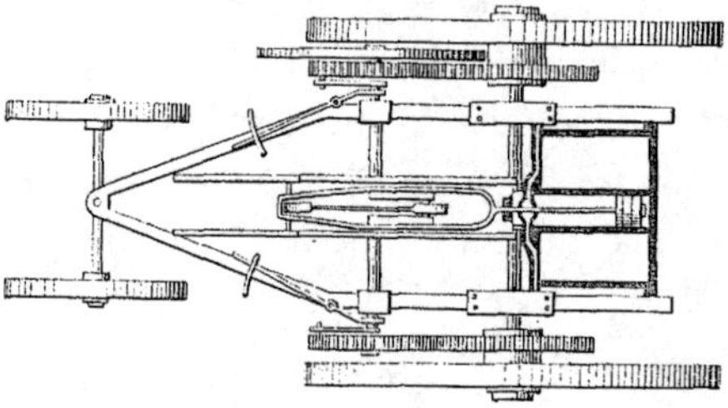

Fig. 14.

Road and Euston Square. One day they started about four o'clock in the morning, and went along Tottenham Court Road, and City Road, and on for four or five miles, travelling at the rate of eight or ten miles an hour. Vivian was steering, and Trevithick was stoking, and they talked of a trip to Cornwall on the way, but on account of careless steering the locomotive ran too close to a wall, and tore down six or seven yards of palings, to the great annoyance of the owner. They at once returned to the coach factory after this mishap. On one occasion, this steam coach ran through Oxford street at a good speed, amid much cheering; no horses or vehicles were allowed on the road during the trial. Trevithick spent nearly three years on road locomotive experiments in Cornwall, and through the streets of London. These trials emptied the pockets of the inventor and his partner, consequently the coach was removed, and the engine was sold for driving a hoop rolling mill, which it continued to do successfully for many years. Trevithick afterwards employed his energies in the construction of locomotives to run on railways, with the very best results. Respecting Trevithick's last days, Smiles says: "The year after the date of his last patent for using superheated steam, he was living at the little village of Dartford, in Kent, employed on some of his inventions in the workshop of Mr. John Hall, when he was seized by the illness of which he died on the 22nd April, 1833. He was lodging at the time at the Bull Inn; but at his death it was found that he had not only outlived all his earnings, but was in debt to the innkeeper. He would, therefore, probably have been buried at the expense of the parish but for his shopmates at Hall's factory, who raised a sum sufficient to give the 'great inventor' a decent burial, and they followed his remains to the grave in Dartford churchyard, where he lies without a stone to mark his resting place. Such was the end of one of the greatest mechanical benefactors of our country."

DUMBELL.—In 1808, John Dumbell patented a peculiar engine, which, among other purposes, was intended for drawing carriages and waggons on common roads. The steam acted on a series of vanes, or fliers, within a cylinder, "like the sails of a windmill," causing them to rotate together with the shaft to which they were fixed. The motion of this shaft was transmitted by suitable gearing to the driving wheels of the carriage. The inventor proposed to raise steam by permitting water to drop upon a metal plate, kept at an intense heat by a fire, which was forced by a pair of bellows. This was a rotary engine. Numerous inventors, Watt among the rest, have tried to apply the steam wheel, as they termed the rotary engine, for the purpose of driving a steam carriage; but none of them accomplished their object.

PRATT.—Major Pratt, in 1810, proposed to propel carriages by means of any motive power mounted upon the carriage frame. Endless chains working over pulleys had arms pivoted to them; these arms, at the ends, carried spikes. The motion of the chains drove the spikes into the ground, and caused the carriage to be slowly driven along.

STEVENS.—John Stevens, of American steam boat fame, in 1812, published a pamphlet urging that railroads and steam carriages should be preferred to canals and canal boats. He asserted his belief that a speed of forty or fifty miles an hour might be obtained, but that probably in practice only thirty miles an hour would be the speed. The speeds given refer to railroads only.

PALMER.—The endless railway was first proposed by Edgeworth, in 1770, and, judging by the patent records, it is destined to be revived again, at repeated intervals, till the end of time. W. Palmer, in 1812, suggested the use of endless chains or rollers, as substitutes for the wheels of carriages.

TINDALL.—In 1814, Thomas Tindall, of Scarborough, patented a steam engine intended for driving all sorts of carriages for the conveyance of passengers, for ploughing land, mowing grass and corn, or for working thrashing machines. The carriage was supported on three wheels, the front one being the steering wheel. On the hinder part of the carriage was mounted a steam engine of most peculiar construction, driving by means of spur gearing, four pushers or legs, which, coming in contact with the ground, drove the carriage forward. The engine could also be made to act upon the two hind wheels for ascending hills, or for drawing heavy loads. It was also proposed to assist the engine by a species of windmill, driven partly by the action of the wind, and partly by the exhaust steam from the engine.

BRUNTON.—Trevithick, as far back as 1800, requested Mr. Gilbert to come with him, and witness the fact that carriages could be propelled by causing their wheels to turn round; and the two friends removed the horse from the carriage when on a stiff hill, and by exerting their strength on the spokes of the wheels, caused the carriage to progress. Furthermore, Trevithick proved, by repeated experiments carried out with the steam carriages we have described, that the adhesion between a pair of good driving wheels and a good road, was quite sufficient to propel a carriage eight miles an hour on the level road, and three or four miles up a steep hill. Yet W. Brunton, of Pentrich, in 1813, obtained a patent for the "mechanical traveller," shown by the illustration herewith (see Fig. 15), for overcoming the supposed difficulty of insufficient adhesion. Brunton's invention was an ingenious method of accomplishing the locomotion of the engine, on an ordinary road, without the aid of the adhesion of the wheels. It consists of a curious combination of levers, the action of which nearly resembles that of the legs of a man in walking, whose feet are alternatively made to press against the ground, and in such a manner as to adapt themselves to the various inequalities of the surface. Fig. 15 shows a side view of the engine. The boiler was of the cylindrical type, with a flue passing through it. The cylinder was placed on one side of the boiler; the piston rod, projecting out behind horizontally, was attached to the leg A B, at A, and to the reciprocating lever A C, which is fixed at C

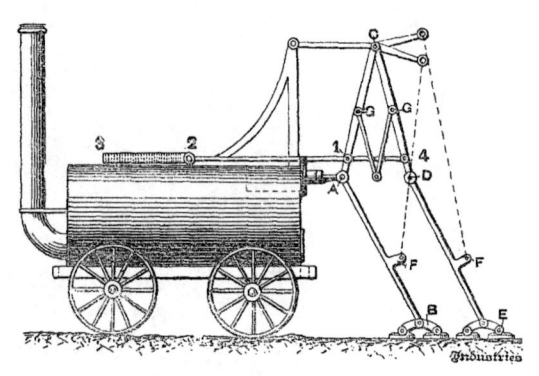

Fig. 15.

At the lower extremity of the leg, feet were attached, by a joint at B. These feet, to lay a firm hold upon the ground, were furnished with short prongs, which prevented them from slipping, and were sufficiently broad to prevent them injuring the road. By referring to Fig. 15, it will be seen that when the piston rod moved outwards it would push the leg A from it, in a direction parallel to the line of the cylinder; but as the leg A B is prevented from moving backwards by the end B being firmly fixed upon the ground, the reaction is thrown upon the carriage, causing it to move forward until the piston reaches the end of its stroke. Upon the reciprocating lever A C is fixed, at 1, a rod 1 2 3, sliding horizontally backwards and forwards upon the top of the boiler. At the end of this rod 2 3 is fixed a rack, which works into a spur pinion lying horizontally upon the top of the boiler. On the opposite side of this spur wheel is a similar rack, and as the wheel turns round the racks are moved in opposite directions, as one leg is forced outwards the other leg is drawn towards the engine ready for a fresh stroke. Whenever, therefore, the piston is at the end of the stroke, and one of the legs is no longer of use to propel the engine forward, the other, immediately on the motion of the piston being changed, is ready, in its turn, to act as a fulcrum for the action of the moving power, to secure the continual progressive motion of the engine. The feet are raised from the ground during the return of the legs towards the engine, by straps fastened at F F, and passing over friction sheaves described in the patent specification. The inventor gives an account of an experiment made with one of these engines, from which we quote the following particulars.* The boiler was of wrought iron, 3ft. in diameter, and 5½ft. long ; the stroke of the piston was 2ft., the length of the step was 26in. The machine only travelled at the rate of two miles an hour, and exerted the power of six horses in hauling a load. The motion must have been very jerky. Brunton's "mechanical traveller" may be regarded as a specimen of great ingenuity, but it could never have worked satisfactorily.

REYNOLDS.—Joseph Reynolds, in 1816, took out a patent for a steam locomotive, to be used for the conveyance

* "Repertory of Arts," Vol. XXIV.

of passengers on ordinary roads, or for drawing agricultural implements. The chief point of novelty in this engine consisted of the mode of suspending the boiler on trunnions, and arranging the steam pipes so that the boiler might always be kept in a horizontal position when the engine was travelling up or down hill. The locomotive was carried at the front end by a single steering wheel, the greater part of the weight being sustained by the two broad driving wheels or rollers. Motion was communicated to each driving wheel separately by spur gearing from the engine crank shaft, and by an arrangement of bevel gearing the driving wheels could be made to revolve in opposite directions for turning in a very small space. The double cylinder engine was bolted to wood framing above the boiler, and the motion of the pistons was communicated through heavy wood connecting rods to the crank shaft. Two or more travelling speeds were provided, and suitable clutch gear for throwing any of the pinions out of gear; brakes were also fitted to the driving wheels.

GORDON.—As far back as 1819, David Gordon, in conjunction with W. Murdock, of Soho, made calculations and carried out experiments for the purpose of using compressed air as a motive power for propelling common road locomotives. Gordon was the inventor of a portable gas apparatus, and impressed with the great importance of locomotion, originated a society of gentlemen, with the ultimate intention of forming a company for the purpose of running a mail coach and other carriages by means of a high pressure engine, or of a gas vacuum or pneumatic engine, supplied with portable gas. Alexander Gordon (son of D. Gordon), in his interesting work on "Elemental Locomotion," states that "the committee of the society had only

THE PERIOD OF EXPERIMENT. 63

a limited sum at their disposal, nor were there to be more funds until a carriage had been propelled for a considerable distance at the rate of ten miles an hour." David Gordon saw that he had been too sanguine about the gas vacuum, and tried to prevail upon the committee to make use of a steam engine, but evidently without success. He then appears to have turned his attention to steam locomotion, for in August, 1821, he took out a patent for improvements in wheel carriages. Fig. 16 represents the type of locomotive proposed by Gordon; it consists of a small steam engine, similar in design to Trevithick's, placed inside a drum 9ft. in

Fig. 16.

diam. and 5ft. wide. The internal circumference of this drum was provided with a series of spur rings, into which were made to gear the spur wheels of a locomotive steam engine. The use of very large wheels to form an endless railway for common road locomotives has been repeatedly suggested since D. Gordon's specification was published. In referring to this engine, A. Gordon states that he had heard of a large drum having been used with advantage for the transport of heavy goods over a swamp in South America. A quantity of iron plates, which were too heavy to be carried otherwise, were rolled into the shape of a large cylinder and riveted together; the inside of the roller was packed with the materials to be conveyed, and as the party advanced this huge cylinder was rolled along with them. Having arrived at their destination, the rivets were cut off, and the plates were applied to their intended purpose. We need hardly remark that an engine made upon Gordon's plan would not injure the roads, but would rather improve them. In 1824, D. Gordon patented a steam carriage similar in construction to the illustration (Fig. 17), which shows a side view of the carriage as actually made. It was mounted upon three wheels, each of which had a separate axle. The single wheel in the front

Fig. 17.

gave an increased facility in turning. In the body of the carriage, connected with the piston rods of the double cylinder engine, was a six-throw crank, to which the propellers were attached, and by the revolution of the crank were successively forced against the ground in a backward direction, then drawn up again. The rods were made of tubes for the sake of lightness, and at the ends of these rods feet are shown, which were made rough on the under side for giving sufficient adhesion to the surface, without digging it up. In order to turn a corner it was only necessary to raise the three propellers on one side and allow the others to act, this being readily effected. Fig. 18 shows the action of the propellers very clearly. The large crank on the left hand shows, at A, C, E. the three throws of the driving crank on one side of the carriage, and the small crank on the right, the same of the lifting crank. The figures, B, D, F show the situation of the three throws on the opposite side of the carriage. It will now be seen that the propeller marked A is, by the position of the crank, just being lifted from the ground, while that marked B,

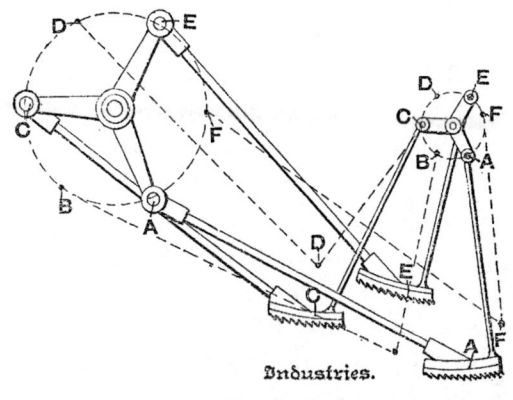

Fig. 18.

which is on the opposite side, would, had the foot been shown, be just commencing its operation; then, C on the near side falls into the position of B and A; then, D on the other side follows, and in like manner E and F. It may be remarked that the lifting rods were hollow, and each had a small solid rod in its interior pressed out by a spiral spring, so that these lifting rods were lengthened when the feet got into a hollow place in the road, and shortened if the feet struck on a stone or other projection between the track of the wheels. In the experiments carried out with this locomotive on the common road the mode of propulsion answered fairly well, but the speed obtained was not satisfactory. In his first trials, Gordon found the power insufficient. He afterwards fitted one of Gurney's light boilers in the hinder part of the carriage, though even after this improvement had been added the experiments were disappointing, and Gordon was convinced that the application of the power to the wheels was the proper mode of propulsion, and, consequently, his project was abandoned, after six or seven years had been spent in inventing, constructing, and carrying out experiments with four distinct carriages.

GRIFFITHS.—In 1821, Julius Griffiths, of Brompton, patented a steam carriage, intended to be used expressly for the conveyance of passengers on the highway; the specification states that the ideas therein expressed were partly communicated by foreigners. The carriage was constructed by the celebrated Joseph Bramah, and although no public trial appears to have been undertaken, yet the carriage was repeatedly tested in Bramah's yard, where it remained for three or four years. It failed to realise the expectations of its promoters on account of a very defective boiler; but A. Gordon says: * "The engines, pumps, and connections were all in the best style of mechanical execution, and had Mr. Griffiths' boiler been of such a kind as to generate regularly the required quantity of steam, a perfect steam carriage must have been the consequence." While the locomotive remained in the maker's yard it was inspected by the many steam carriage schemers of the period, and hence some of the details and mechanical combinations served as useful hints to succeeding mechanicians in similar undertakings, and were afterwards embodied in their steam carriage proposals. Fig. 19 gives a side elevation of Griffiths' carriage which has been termed the "first steam coach constructed in this country

* "Elemental Locomotion," 1832.

THE PERIOD OF EXPERIMENT.   67

expressly for the conveyance of passengers on common roads." The two vertical working steam cylinders, the boiler, condenser, and other details were suspended to a wood framing at the hinder part of the carriage, as shown in the engraving. The coach was of the ordinary kind, suspended on springs. The engine driver had a seat behind, where he could attend to the stoking, and start and stop the engine. The boiler consisted of a series of horizontal water tubes, 1½in. diam. and 2ft. long, the flanges of which at each end were bolted to the vertical tubes forming the sides of the furnace; it was, however, deficient in steam room, and primed badly. A small water tank was strapped to the wood framing in front of the

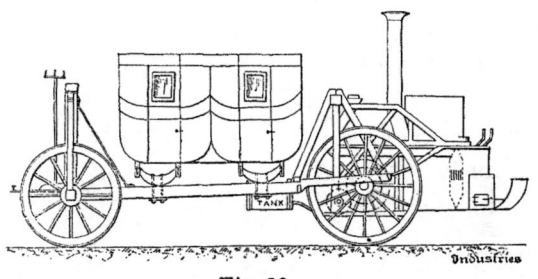

Fig. 19.

driving wheels, and a force pump supplied the boiler with water therefrom. The steam, after it had done its work in the cylinder, was conducted to an air condenser consisting of a number of flattened thin metal tubes exposed to the cooling influences of the air. "The power of the engines was communicated from the piston rods to the driving wheels of the carriage through the means of sweep rods, the lower ends of which are provided with driving pinions and detents, which operate upon toothed gear fixed to the hind carriage axle." The object of this mechanism, which is of foreign invention, says Luke Hebert, "is to keep the driving pinions always in gear with the toothed wheels, however the engine and other machinery may vibrate, or the wheels be jolted upon uneven ground." In order that the boiler, engine, and other working parts might not suffer from the violent vibrations experienced when passing over rough roads, they were suspended to the wood framing by chain slings, having strong spiral springs introduced between them, as shown in the drawing. The carriage was steered by means of the usual mechanism, shown at the forward end.

BROWN.—In 1823 Samuel Brown patented a locomotive for common roads, and three years later fitted up a carriage propelled by his patent gas vacuum engine, on which he ascended Shooter's Hill to the satisfaction of numerous spectators. The Canal Gas Engine Company was formed for applying this engine to the propulsion of vessels either on canals or navigable rivers, and several experiments were carried out on the Thames; but the great expense of working such an engine caused it to be laid aside, and the company was soon dissolved.

BURSTALL AND HILL.—In 1824 a patent was granted to T. Burstall, of Edinburgh, and J. Hill, of London, for a locomotive engine. A side elevation of the carriage as actually constructed is shown in Fig. 20. The boiler was of peculiar design, when seen in section; the shell was formed of cast iron, and the inner portion of the boiler consisted of shallow trays or receptacles for containing a small quantity of water to be converted into steam. The object of the patentees was to make the boiler a store of heat. "They proposed to heat it from 250°F. to 600°F., and by keeping the water in a separate vessel, and only applying it to the boiler when steam was wanted, they accomplished that desideratum of making just such a

BURSTALL & HILL'S STEAM COACH, 1824.

Fig. 20.

quantity of steam as was wanted ; so that when going down hill all the steam and heat might be saved, to be accumulated and given out again at the first hill or piece of rough road, when, more being wanted, more will be expended."* The boiler was very much larger than it appears in the side view (Fig. 20), as the furnace extended across the full width of the carriage. The gases were conducted through a square flue running round the outside of the boiler shell, and leading into the large chimney which rose from the centre. The boiler was suspended on springs, and the pipes for conducting steam from the boiler to the cylinders were made of copper, bent in the form shown, so that the vibration of the boiler and the rest of the machinery should not break any of the joints. The feed water cistern shown beneath the longitudinal driving shaft was made of sheet copper, air tight, and strong enough to sustain a pressure of 6olb. per square inch. Two air pumps were worked by the beams for forcing air into the reservoir, so that this pressure would drive the water through a convenient pipe from the tank into the boiler, at such time and in such quantity as might be required. The two cylinders were each 7in. diam., steam being admitted above and below the pistons by suitable valves. The stroke was 12in., and when working at high pressure the engine was said to indicate 10 h.p. The exhaust steam was let off into an intermediate receiver, so that the noise could be lessened. The two vibrating beams were connected at one end to the piston rod, and at the other end to rocking standards carried on the top of the boiler shell. At about a fourth of the length of the beam were the two connecting rods driving the cranks on the hind axle, the cranks being

* "Edinburgh Philosophical Journal."

THE PERIOD OF EXPERIMENT.

placed at an angle of 90° from each other. All the wheels of the carriages were used as drivers when ascending hills, drawing other vehicles behind, or traversing rough roads. This was effected by means of a pair of mitre wheels placed on the hind axle driving a longitudinal shaft, which in its turn drove the front axle through bevel wheels (shown more clearly in Fig. 21) whether the front axle was in the lock or not. None of the wheels were keyed direct to the axles, but the bosses of the driving wheels were provided with ratchets, discs, and spring pawls inside; these ratchet wheels drove the travelling wheels when the axles revolved, and at the same time allowed the outer wheel, when the

Fig. 21.

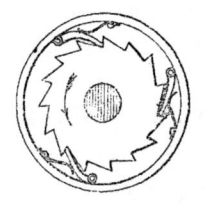

Fig. 22.

carriage described a curve to travel faster than the inner one, and still be ready to receive the impulse of the engine as soon as the carriage returned to a straight course. Fig. 22 shows the ratchet wheel and spring pawl. When it was necessary to back the coach the pawl and the ratchet wheel had to be locked; but Burstall and Hill patented another method of performing the same operation. The wheel naves were cast with a recess in the centre, in which was fitted a double bevel clutch, the inside of the boss being cast to correspond; and by a simple lever these clutches were made to drive when going forward or backward. Considerable ingenuity was displayed in the details and arrangement of this carriage; but throughout the repeated experiments made with it in London, no real success was achieved, in spite of many lengthy alterations being carried into effect between the trials. The utmost speed attained was from three to four miles an hour on an enclosed piece of ground, failure being altogether due to the boiler, upon which much money was expended. Mr. Hebert, the editor of the "Repertory of Patent Inventions," when referring to Burstall and Hill's carriage, suggested that the great impediment to the application of steam carriages to common roads was their enormous weight. This averaged about 8 tons, and if the usual loads put on an eight-horse waggon be added to this no common road then made would support such a weight; and he hinted that it would be well for the patentees of this carriage to contrive means for lessening the weight by having the engine and boiler quite distinct from the carriages used for conveying the passengers or merchandise, like the plan pursued on the railways. Passengers were afraid of sitting near the boiler, and in warm weather the great heat was very objectionable; whilst the jar of the machinery working close to the inmates of the coach was an additional reason for making the engine and boiler form an "iron horse," to be connected to the carriages by couplings, and not to be self contained, like Burstall and Hill's steam carriage. A further patent for improvements in steam carriages was obtained by Burstall and Hill in 1826, and in the following year they completed a carriage embodying these improvements, which was exhibited in Leith and Edinburgh, and in front of the Bethlehem Hospital, London; but it was soon abandoned, and no particulars appear to have been chronicled respecting the design of this engine. Burstall and Hill, probably acting upon the advice given by Mr. Hebert, constructed in 1827 a working model of a steam carriage, which is shown in Fig. 23. In this a simple vertical boiler was used on a separate carriage placed behind the engine and the coach. The model was one-fourth full size, and was exhibited at work in Edinburgh, and afterwards in London. The length of the model was 5ft. 6 in., and its height 1ft. 10in., corresponding to a length of 22ft., and a height of 7ft. 3in. The boiler was of the ordinary vertical conical type, the inside fire box and the shell being made of copper; it was intended for a working pressure of 25lb. of steam per square inch, and was tested up to a very high pressure. A

Fig. 23.

double cylinder high pressure engine was placed in the hind part of the body of the coach; the cylinders were 3in. diam., with a stroke of 3in. The full sized engine at 25lb. pressure would develop 10 h.p. The cylinders were placed vertically, the piston rods working in guides, and the connecting rods actuated the cranks on the ends of the main driving axle outside the wheels. An ordinary force pump was used in this case for feeding the boiler, and the exhaust steam was led into the chimney to create a blast. The coach was of the usual construction, intended to carry six passengers inside and twelve passengers outside. The weight of the full sized carriage without fuel and water would be about 3 tons, and this weight equally distributed on six broad tyred wheels would not injure the roads. The carriage was steered in a simple manner. "The proportion of the locomotive was accurately preserved in all its parts, and exhibited a faithful representative of the carriage to be used on the road (if successful), so that it afforded a measure (though not a very correct one) of possible performances."* The model was made to travel round a circular and very uneven platform of deals 17ft. diam., with a speed equal to seven or eight miles an hour. An incline 18ft. long, rising one in eighteen, was fixed, and the engine ran up with ease and rapidity. The steam carriage was subjected to the roughest usage by being run over tools of various descriptions laid in the way, and it was asserted that this model ran in the space of eight days 250 miles without needing the glands to be re-packed or any repairs. It is to be regretted that this promising model never gave rise to actual working engines, as these would doubtless have succeeded. Had Burstall and Hill's first carriage been fitted with a more simple type of boiler it would most likely have also done well.

JAMES.—W. H. James, of Thavies Inn, Holborn, whom Luke Hebert designates a gentleman of superior mechanical talents, in 1823 patented a tubular steam boiler designed expressly for road locomotives. The year following he introduced a steam carriage which possessed several points of novelty. Fig. 24 gives a plan, Fig. 25 a sectional end elevation, and Fig. 26 a sectional side view. It will be observed that a double cylinder engine is employed to drive each hind wheel. These cylinders are small in diameter,

* Galloway on the Steam Engine.

and the pistons were worked by steam at a pressure of 200lb. per square inch. The object of the patentee in employing separate engines to each driver was to give each wheel an independent motion, so that the power and speed might be varied at pleasure for turning corners, the outer wheel travelling over a much greater space than the inner wheel. By a suitable arrangement of cocks in the steam pipe, to which the front axle was connected, the amount of steam

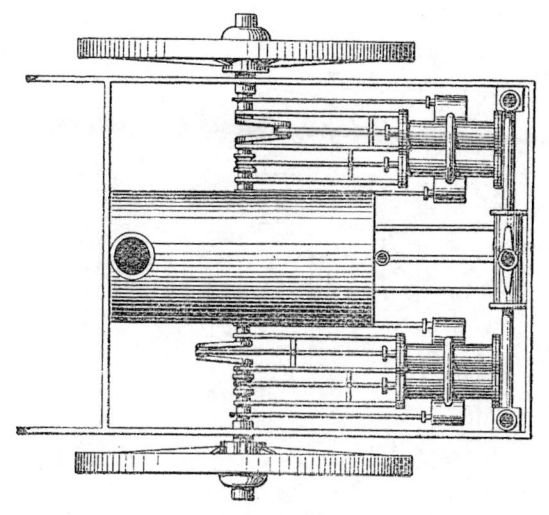

Fig. 24.

admitted to each engine was automatically controlled. When the front wheels were in such a position that the carriage proceeded in a straight line an equal amount of steam was admitted to each pair of cylinders, but when the front wheel was in the lock the engine driving the outer wheel received a greater amount of steam, and thus developed more power and travelled faster than the inner wheel. Hebert describes it as being "so efficient that the carriage could be made to describe every variety of curve; he has seen it repeatedly make turns of less than 10ft. radius."* The whole of the machinery was very ingeniously mounted upon laminated carriage springs, the advantages of which are well understood. This was effected by causing the engines and their frame work to vibrate altogether upon the crank shaft as a centre, at the same time connecting these engines to the boiler by means of hollow axles moving in stuffing boxes. The arrangement of springs, the engine frames, and steam

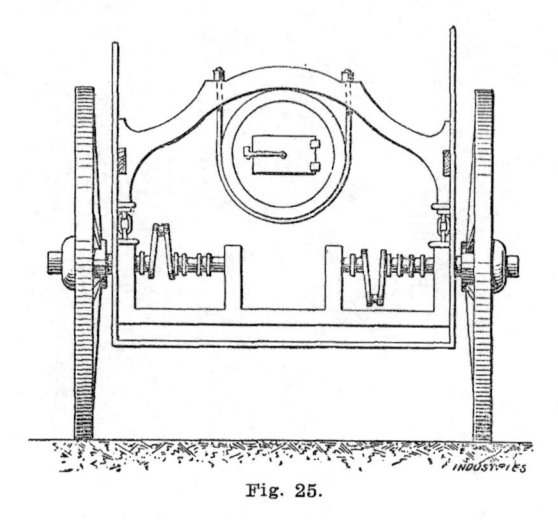

Fig. 25.

pipe may be seen from Fig. 26. Each engine has two cylinders of small diameter and long stroke; to these separate engines steam is supplied from the boiler by means of the main pipe, which moves through steam tight stuffing boxes to the slide valve boxes by small pipes, as shown. The slide valves are worked by eccentrics on the engine shafts in the usual manner, and the exhaust steam conducted into the chimney. It will be noticed that James adopted the wise course of making the locomotive quite distinct from

* Register of Arts." 1829.

THE PERIOD OF EXPERIMENT.   77

the carriage intended for passengers, and on the whole the engine we have described was certainly a decided advance, compared with any of its immediate predecessors, and judging from the drawings it should have proved a success. Owing to pecuniary difficulties he was not able to build a carriage upon the lines we have described, but he was in the meantime busily engaged in designing high pressure tubular boilers for road locomotives, &c. At length, however, Sir James Anderson, Bart., of Buttevant Castle, Ireland, connected himself with James in the construction of steam

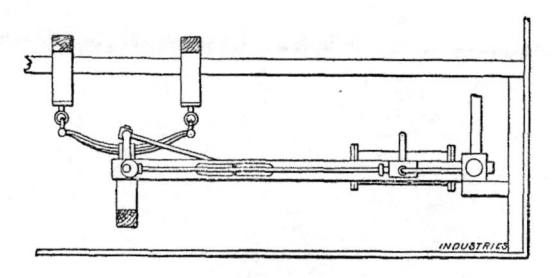

Fig. 26.

carriages, and in March, 1829, they carried out some experiments with a steam coach as shown by Fig. 27, the engine of which was very similar to that illustrated in Figs. 24, 25, and 26. This weighed nearly three tons, and the first trials were made round a circle of 160ft. in diameter; in every experiment some defect being discovered and altered. At last the engine was brought out and loaded with fifteen passengers, and was propelled several miles on a rough gravel road across Epping Forest, with a speed varying from twelve to fifteen miles an hour. Steam was supplied by two tubular boilers, each forming a hollow cylinder, 4ft. 6in. long. It should here be observed, says Hebert,[†]

[†] In the Appendix to Galloway's Treatise on the Steam Engine, 1830.

JAMES' STEAM COACH, 1829.

Fig 27

"that the tubes of which the boilers were composed were common gas tubes, one of which split, thus letting the water out of one of the boilers and extinguishing its fire. Under these circumstances, with only one boiler in operation, the carriage returned home at the rate of about seven miles an hour, carrying more than twenty passengers, at one period indeed, it is said, a much greater number; showing that sufficient steam can be generated in such a boiler to be equal to the propulsion of between 5 tons and 6 tons weight. In consequence of this flattering

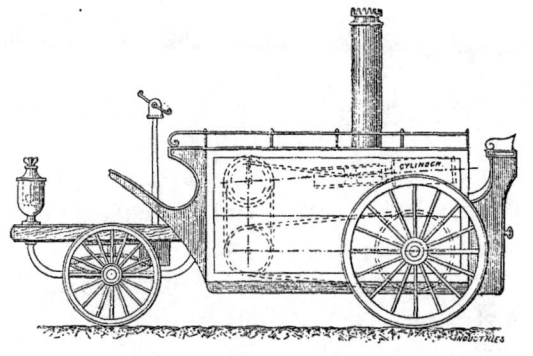

Fig. 28.

demonstration that the most brilliant success was attainable, the proprietors dismantled the carriage and commenced the construction of superior tubular boilers with much stronger tubes."

Shortly after the trials conducted by Sir J. Anderson and W. H. James across Epping Forest with the steam carriage made in accordance with James' patents of 1824 and 1825, they commenced to build another steam carriage, which was ready for use in November, 1829. Fig. 28 may be taken as a representation of the outside appearance of this fresh venture, but the internal details were of different design. This engine was not intended to carry any passengers, but to be employed for drawing carriages behind. Four tubular boilers were used, the total number of tubes being nearly two hundred. These boilers were enclosed in a space 4ft. wide, 3ft. long, and 2ft. deep. The steam from each boiler was conducted into one main steam pipe $1\frac{1}{2}$in. in diameter, and the communication from any one of the boilers could be cut off in case of leakage. Four cylinders, each $2\frac{1}{4}$in. bore and 9in. stroke, were arranged vertically in the hind part of the locomotive, and two of them acted upon each crank shaft as before, giving a separate motion to each driving wheel. The exhaust steam was conducted through two copper tanks for heating the feed water to a high temperature, and thence passed to the chimney. The steering gear consisted of an external pillar containing a vertical shaft, at the upper end of which small bevel gearing was used, giving motion to the vertical shaft, whose bottom end carried a pinion gearing into a sector attached to the fore axle. A large lamp is shown in the illustration for lighting the road for the steersman. The motion of the crank shafts was in this case communicated to the separate axles of the driving wheels by spur gearing with two speeds, changeable at will and as the state of the road required. In numerous experiments made with this carriage, in which Hebert had an opportunity of riding, he says, "the greatest speed obtained upon a level on a very indifferent road was at the rate of fifteen miles an hour," and it never ran more than three or four miles without breaking some of the steam joints, the workmanship of the day not being good enough for a steam pressure of 300lb. to the square inch. We quote the following from the *Mechanics' Magazine* of the 14th November, 1829: "A series of interesting experiments were made throughout the whole of yesterday with a new steam carriage belonging to Sir James Anderson, Bart., and W. H. James, Esq., on the Vauxhall, Kensington, and Clapham roads, with the view of ascertaining the practical advantages of some perfectly novel apparatus attached to the engines, the results of which were so satisfactory that the proprietors intend immediately establishing several stage coaches on the principle. The writer was favoured with a ride during the last experiment, when the machine proceeded from Vauxhall Bridge to the Swan at Clapham, a distance of two and a half miles, which was run at the rate of fifteen miles an hour. From what I had the pleasure of witnessing, I am confident that this carriage is far superior to every other locomotive carriage hitherto brought before the public, and that she will easily perform fifteen miles an hour throughout a long journey, The body of the carriage, if not elegant, is neat, being the figure of a parallelogram. It is a very small and compact machine, and runs upon four wheels."

W. H. James patented another steam carriage in August 1832. An illustration of this is given in Fig. 28, which shows a wide departure in the working parts from his earlier engines. The two high pressure cylinders were fixed near the top corner over the boiler, and the pistons actuated, by means of long connecting rods, a crank shaft placed near the front top corner of the carriage. The shaft carried three chain wheels of different diameters, and these communicated motion to three similar wheels placed on a counter shaft fixed directly under the crank shaft. The chain wheels being of different diameters three distinct travelling speeds were provided for, since any pair of chain wheels could readily be put into gear by means of clutches actuated by foot levers placed beneath the steersman. It will be seen from the illustration that motion was communicated from the counter shaft to the driving axle by means of a pitch chain and suitable chain wheels. "A bar on the fore carriage was connected by chains to rods working cams, which threw in and out of gear ratchets on the bosses of two of the chain wheels communicating motion to the driving wheels The effect of this arrangement was that on the fore carriage being turned from side to side to steer the engine, the driving wheels on one side were thrown out of gear." The fire was blown by bellows placed under the steersman's seat, worked by the engine. The boilers were of the water tube type, and very complicated. A double acting force pump was arranged to produce a circulation of water through the boiler. The locomotive carried a large water tank communicating by tubes with the boiler. A surface condenser was placed in the bottom of the carriage. The details of this locomotive were not nearly so well arranged as in those previously made by James. The method of disconnecting the driving wheels when turning corners could not have worked satisfactorily. We are not sure that a carriage was ever built in accordance with the patented plans of 1832, as Sir James Anderson appears to have fallen into pecuniary difficulties about this time, according to Maceroni.

NEVILLE.—In 1823, James Neville, of Shad Thames, London, took out a patent for a steam boiler intended for steam carriages, and in 1827 patented an improved road locomotive, the chief object of which appears to have been to provide the periphery of the driving wheels with spikes to prevent them from slipping. One of the first spring wheels is embodied in Neville's specification. He says: "When the steam carriage is intended to run up very steep inclines, I attach, if necessary, plates of elastic steel." These plates (shown in Fig. 29) were 18in. long, and made rough on their outside surface by projecting steel studs. They were made the full width of the tyre, and fixed by countersunk bolts, so that the springs when not compressed formed tangents to the circumference of the wheel, as shown in the illustration.

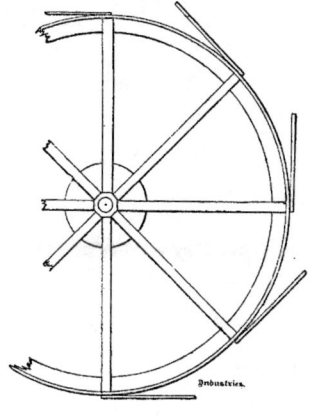

Fig. 29.

Their elasticity enabled them to assume the circular form of the tyre, as they ascended with the wheel from the ground; at the same time their extended surface occasioned greater resistance, and prevented the wheels from slipping when ascending steep hills. Fig. 30 shows a plan of Neville's steam carriage, from which it will be seen that two oscillating cylinders were placed horizontally beneath the carriage: the piston rod ends were attached to the double cranked axle, no connecting rods or guide bars being necessary. The carriage could be propelled by the engines working directly, or if hills had to be surmounted, rough roads to be traversed, or heavy loads to be hauled, a countershaft was provided with spur gearing to transmit the power, suitable clutches being provided for quickly bringing this gearing into action when required. The boiler employed is one patented in 1826.

NEVILLE'S STEAM CARRIAGE, 1827.

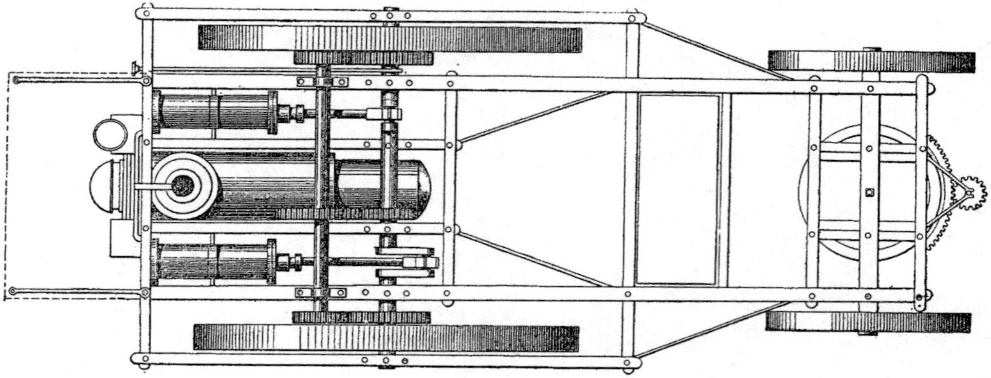

Fig. 30.

SEAWARD.—In 1825 J. and S. Seaward, of the City Canal Iron Works, London, patented a method of propelling road locomotives, by means of a wheel or wheels connected by a swinging frame to the crank shaft of a steam engine, as shown in the sketch Fig. 31. The propelling wheels could

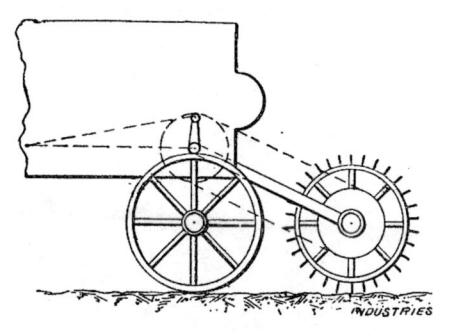

Fig. 31.

also be fixed in circular grooves, so that they might rise and fall to accommodate themselves to the inequalities of the road over which they travelled. The weight of the engine was to be carried upon separate wheels. The tyres of these propellers had projecting teeth or indented surfaces, to give them a hold on the ground or cause them to "bite" and prevent them from slipping. Seaward's propelling wheels have been repeatedly revived, and patented in modified forms. These latter schemes will be noticed in due course, but we may state in passing that there cannot be two opinions respecting the injurious effect of such spiked wheels upon the roads, seeing that rotary diggers are exactly similar in principle.

PARKER.—T. W. Parker, of Edgar County, Illinois, in 1825, made a large working model of a steam carriage of simple construction, and very light. It was mounted upon three wheels, the leading wheel as well as the two hind wheels, 8ft. in diameter, being driven by a double cylinder engine. The carriage was examined by many persons, and no fault appears to have been found with its action.

ANDREWS.—F. Andrews, of Stamford Rivers, in Essex, patented some improvements in steam carriages in 1826. He appears to have been the inventor of what is termed

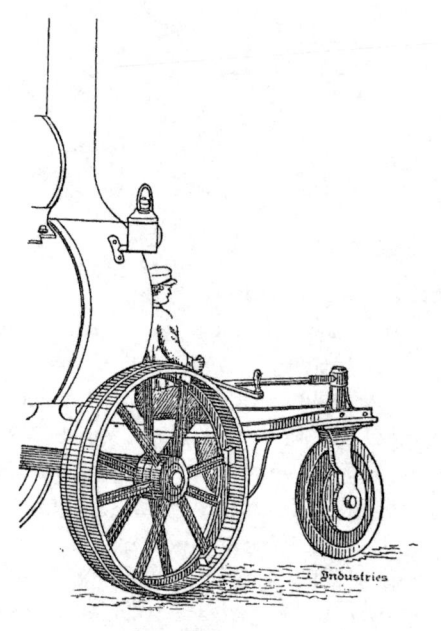

Fig. 32.

the " pilot " steering wheel, a plan which Gurney appropriated two years after the patent was sealed. It will be noticed that Gurney's steam carriage of 1828 was fitted with pilot steering. Thomas Aveling re-patented this form of steerage about 1860, and there are traction engines in existence at the present time fitted with fore carriages

THE PERIOD OF EXPERIMENT. 87

of such design. The pilot steerage consisted of a single disc running between two shafts, which were connected to the leading axle, and as the disc was turned by the steersman's lever to the right or left hand, the front wheels followed in the same direction. Fig. 32 shows clearly this type of steerage. Another novelty claimed by Andrews was the use of cylinders working on trunnions, the piston rods of which acted directly on the crank shaft. The oscillating cylinders were placed in a horizontal position, so that they were quite unaffected by the motion of the springs. These cylinders were used by

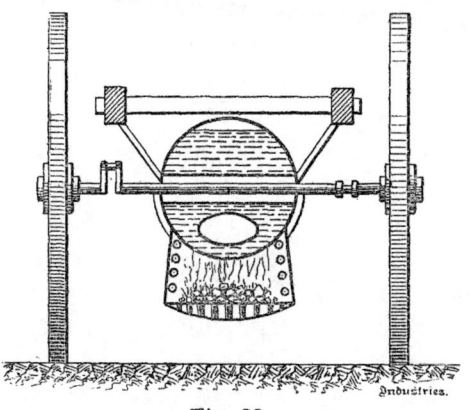

Fig. 33.

Neville in his steam carriage (see Fig. 30), but Andrews was the designer. Neville says in his patent specification: " I do not confine myself to any particular form of steam engine, neither do I think a detailed description of such engines necessary." Fig. 33 gives a sectional view of Andrews's steam carriage; the furnace is shown at the bottom, and the products of combustion pass along underneath the bottom of the boiler, and return to the front of the boiler through the flue, thence into the chimney, which is not shown. The cranked axle was encased in a tube passing through the water space of the boiler, the opening being large enough to allow the crank to pass through. The boiler was hung upon framing, and supported on springs. Mr. A. F. Yarrow, in his paper on "Steam Carriages," 1862, says: "Andrews's steam carriage, like many others, proved unsuccessful, owing to the failure of the boiler."

GOUGH.—Nathan Gough, of Salford, is the next inventor to be noticed. He patented some steam carriage arrangements possessing merit, but as no carriages appear to have been built to his plans, we cannot stay to describe them.

HOLLAND.—T. S. Holland took out a patent in 1827 for a curious method of producing a locomotive action for steam carriages.

NASMYTH.—The following account of a steam carriage made by Mr. James Nasmyth is quoted from his autobiography, edited by Dr. Smiles.* "About the year 1827, when I was 19 years old, the subject of steam carriages to run upon common roads occupied considerable attention. Several engineers and mechanical schemers had tried their hands, but as yet no substantial results had come of their attempts to solve the problem. Like others, I tried my hand. Having made a small working model of a steam carriage, I exhibited it before the members of the Scottish Society of Arts. The performance of this active little machine was so gratifying to the Society, that they requested me to construct one of such power as to enable four or six

* James Nasmyth, Engineer, an Autobiography S. Smiles, LL.D., 1885.

persons to be conveyed along the ordinary roads. The members of the Society, in their individual capacity, subscribed £60, which they placed in my hands as the means for carrying out their project. I accordingly set to work at once, and completed the carriage in about four months, when it was exhibited before the members of the Society of Arts. Many successful trials were made with it on the Queensferry Road near Edinburgh. The runs were generally of four or five miles, with a load of eight passengers sitting on benches about 3ft. from the ground. The experiments were continued for nearly three months, to the great satisfaction of the members. I may mention that in my steam carriage I employed the waste steam to create a blast or draught, by discharging it into the short chimney of the boiler at its lowest part; and I found it most effective. I was not at that time aware that George Stephenson and others* had adopted the same method; but it was afterwards gratifying to me to find that I had been correct as regards the important uses of the steam blast in the chimney. In fact, it is to this use of the waste steam that we owe the practical success of the locomotive engine as a tractive power on railways, especially at high speeds. The Society of Arts did not attach any commercial value to my road carriage. It was merely as a matter of experiment that they had invited me to construct it. When it proved successful they made me a present of the entire apparatus. As I was anxious to get on with my studies, and to prepare for the work of practical engineering, I proceeded no further. I broke up the steam carriage, and sold the two small high pressure engines, provided with a strong boiler, for £67, a sum which more than defrayed all the expenses of the

* The blast pipe was introduced by Trevithick in his first steam road locomotive, some years before George Stephenson used it.

construction and working of the machine. James Nasmyth died 6th May, 1890, aged 82 years.

VINEY.—Colonel James Viney, Royal Engineers, patented a boiler in 1829, which was intended for steam carriages. The boiler was arranged with two, three, four, or six concentric hollow cylinders containing water, between which the fire from below passed up. There was shown an annular space for water, and an annular space or flue for the ascending fire, alternately; the water being placed between two fires. It does not appear that this boiler was ever tried.

HARLAND.—Dr. Harland, of Scarborough, in 1827 invented and patented a steam carriage for running on common roads. "A working model of the steam coach was perfected, embracing a multitubular boiler for quickly raising high-pressure steam, with a revolving surface condenser for reducing the steam to water again by means of its exposure to the cold draught of the atmosphere through the interstices of extremely thin laminations of copper plates. The entire machinery, placed under the bottom of the carriage, was borne on springs; the whole being of an elegant form. This model steam carriage ascended with ease the steepest roads. Its success was so complete that Dr. Harland designed a full-sized carriage; but the demands upon his professional skill were so great that he was prevented going further than constructing the pair of engines the wheels, and a part of the boiler, all of which remnants, Dr. Smiles tells us, 'are still preserved as valuable links in the progress of steam locomotion.'"* Dr. Harland had a great love for mechanical pursuits. He spent his leisure time in inventions of many sorts; and, in conjunction with the late

* "Men of Invention and Industry," by S. Smiles, LL.D., 1884.

Sir George Cayley of Brompton, he kept an excellent mechanic constantly at work. Dr. Harland was thrice Mayor of Scarborough, and a Justice of the Peace for the borough. He died in 1866.

RAWE AND BOASE.—In 1830, a patent was issued to Rawe and Boase, of Albany Street, London, for improvements in steam carriages.

CLIVE.—Mr. Clive, of Chell House, Staffordshire, took out a patent in 1830 for improvements in the construction of locomotives, consisting chiefly of two proposals, according to which the driving wheels were to be made from 5ft. to 10ft. diam., according to circumstances, and the throw of the cranks was to be increased to from 18in. to 24in. Clive wrote many articles on steam carriages under the signature of "Saxula" in the *Mechanics' Magazine.* In 1843 he writes as under:—" I am an old common road steam carriage projector, but gave it up as impracticable ten years ago, and I am a warm admirer of Col. Maceroni's inventions. My opinion for years has been, and often so expressed, that it is impossible to build an engine sufficiently strong to run even without a load on a common road, year by year, at the rate of 15 to 20 miles an hour. It would break down. Cold iron at that speed cannot stand the shock of the momentum of a constant fall from stones and ruts of even an inch high."

LEA.—The following letters relating to the work of Lea appeared in the *Mechanics' Magazine* during the latter part of the year 1830: " I beg to mention that a mechanic named Lea, of Hoxton, has constructed a small model of a steam carriage capable of accomplishing everything that a horse can do, on any road or in any season, and it differs in its principle from every steam travelling apparatus yet before the public." (Signed "W. Turner, Old Hoxton.") To this letter "Saxula" replied as follows: "I request Mr. Turner to propose to Lea the following test: Ascertain whether his steam carriage will ascend, even with difficulty, on plain wheels, an incline of one in three. If it fail to do this, and the wheels turn constantly round without moving at all—and it is not the first competitor that has failed—he may be assured that his locomotive, although on a new principle, is not built on the true principles of locomotion."

HEATON.—W. G. & R. Heaton, of Birmingham, built several carriages under a patent taken out in 1830. The mechanism adopted was very complicated, and the patentees acknowledged that none of the separate details were novel. They merely claimed the combination constituting the general structure. Their boiler and steering gear were like those of James's, and their mode of driving resembled Trevithick's. In August, 1833, Messrs. Heaton placed a steam drag on the road between Worcester and Birmingham, but some part of their machinery unfortunately broke in ascending the Lickey Hill. After having the damage repaired, they started again on the same road. "Attached to the engine was a stage coach, carrying fifteen passengers weighing 1 ton 15 cwt. They picked up five more passengers shortly after starting, and arrived at Northfield, a distance of nearly seven miles, in 56 minutes. Having taken in water, they started, and proceeded to ascend the Lickey Hill, a rise of one in nine, and even one in eight in some places; many parts of the hill were very soft, but by putting both wheels in gear they ascended to the summit, 700 yards

in nine minutes. After proceeding to Broomsgrove, the drag and carriage returned, and on descending the steepest part of the hill they proved their power by stopping suddenly. This hill is one of the worst upon any turnpike road in England."*
In 1833 a company was formed in Birmingham for the purpose of constructing and running one of Messrs Heaton's carriages, subject to the condition of keeping up an average speed of ten miles an hour. After repeated trials during 1834 with a new carriage, Messrs. Heaton dissolved their contract with the company, candidly declaring their inability to do more than seven or eight miles an hour.†

The following appeared in the *Birmingham Journal*, 12th April, 1834 :—"We are authorised by 'Heaton's Steam Carriage Company' to state that the results of experiments have not proved satisfactory, and they will call a meeting of

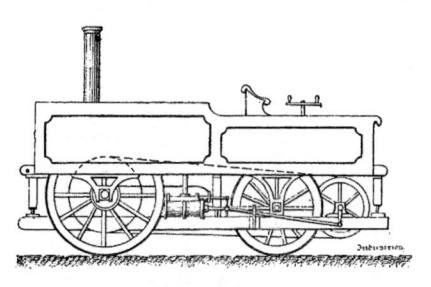

Fig. 34.

the shareholders to take into consideration a letter from Mr. Heaton on the subject. After spending upwards of £2,000 in endeavours to effect steam travelling, Messrs. Heaton honourably retire from the field, stating that the wear and tear were excessive at 10 miles an hour; the carriage was heavy, and wasteful in steam, hence their reasons for giving up."

*Young's " Steam on Common Roads."
†Maceroni's " Steam Power Applied to Common Roads.

We are sorry that Heatons discontinued running their carriages, which did well at slower speeds, as the company they had formed would have accepted the carriages at a speed of eight miles an hour.

NAPIER.—Messrs. Napier, of London and Glasgow, in 1831, took out a patent for a steam locomotive, constructed as shown in Fig. 34. The boiler was of the horizontal type, placed beneath the carriage, to which the two steam cylinders were bolted. The crank shaft, which was placed in front of the steering wheel, had a chain wheel keyed upon it, which communicated motion from the crank shaft to the driving axle as shown. Several chains were to be used if necessary. The boiler and engine were bolted together and secured to the under framing of the carriage, the upper portion being mounted upon spiral springs.

THE PERIOD OF SUCCESSFUL APPLICATION.

SUMMERS AND OGLE.—In 1831, Summers and Ogle built two steam carriages which are noted for having attained extraordinarily high speeds on common roads. Their boilers were similar to those used on American fire engines, and were worked at a pressure of 250lb. per square inch. The fuel used was soft and good coke, which produced no smoke. The first carriage was fitted with two steam cylinders, each 7½in. diam. and 18in. stroke. The carriage was mounted upon three wheels, the drivers being 5ft. 6in. diam. Fig. 35

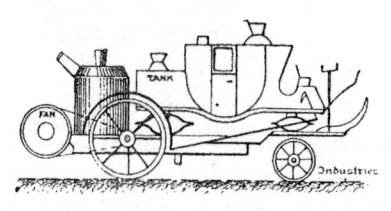

Fig. 35.

gives an outline sketch of this carriage, from which it will be seen that the passengers were placed in the front and middle of the vehicle, the boiler being placed behind the body of the carriage. The exhaust steam was not turned into the chimney, but the fire was blown with a fan driven by the engine. Their second carriage was very similar in construction, but the boiler was transferred from one vehicle to the other, and three steam cylinders were used, each 4in. diam. with 12in. stroke. Summers and Ogle, in their evidence before a committee of the House of Commons, gave the following particulars of their trials. The greatest velocity they ever obtained was thirty-two miles an hour; they went from the turnpike gate at Southampton to the four-mile stone on the London road, a continued elevation, with one slight descent, at the rate of twenty-four and a half miles per hour, loaded with people. Their first steam carriage ran from Cable-street, Wellclose Square, to within two miles and a half of Basingstoke, when the crank shaft broke, and they were obliged to put the whole machine into a barge on the canal and send it back to London. This same machine had previously run in various directions about the streets and outskirts of London. With their improved carriage they went from Southampton to Birmingham, Liverpool and London, with the greatest success. The following is an abbreviated account of one of their trials from the *Saturday Magazine*, October 6th, 1832 : " I have just returned from witnessing the triumph of science in mechanics, by travelling along a hilly and crooked road from Oxford to Birmingham in a steam carriage. This truly wonderful machine is the invention of Captain Ogle, of the Royal Navy, and Mr. Summers, his partner, and is the first and only one that has accomplished so long a journey over chance roads, and without rails. Its rate of travelling may be called twelve miles an hour, but twenty or perhaps thirty down hill if not checked by the brake, a contrivance which places the whole of the machinery under complete control. Away went the splendid vehicle through that beauteous city (Oxford) at the rate of ten miles an hour, which, when clear of the houses, was accelerated to fourteen. Just as the steam carriage was entering the town of Birmingham, the supply of coke being

THE PERIOD OF SUCCESSFUL APPLICATION. 97

exhausted, the steam dropped ; and the good people, on learning the cause, flew to the frame, and dragged it into the inn yard." The carriage weighed 3 tons with coal and water, but without passengers, of which as many as twenty were sometimes carried.

GURNEY.—Many of the steam carriage schemes thus far dealt with, judging at least from appearances, gave little promise of success, the promoters in numerous instances being persons of no mechanical experience. Others deserved to succeed on the score of meritorious design. The arrangement of some of these vehicles presented many valuable features, but owing to structural defects and poor workmanship, even these more promising steam carriages did not work satisfactorily, the boiler being generally one of the causes of failure. We have now, however, arrived at a period in the history of steam locomotion when success was practically achieved ; common road carriages were in successful operation, and many carriages were being built for regular passenger service in various parts of the country. Among those who laboured in the steam carriage industry when success was realised, Sir Goldsworthy Gurney occupies a prominent, though not by any means the foremost place, since he was not so successful as Hancock, and other of his contemporaries. Some authors have ignored all the honest work connected with the introduction of steam locomotion on common roads. that has been performed by the thirty or more labourers whose inventions we have described, and who lived previous to Gurney's day. They have consequently placed Gurney in a too exalted position ; for instance, Dr. Lardner says (*The Steam Engine*): " First and most prominent in the history of the application of steam to the propelling of carriages on turnpike roads stands the name of

Sir Goldsworthy Gurney." As Gurney's inventions have been so often described, it will only be necessary for us to briefly sketch his career. He was born at Trevorgus, in Cornwall, in 1793. In 1822 Gurney stated in a public lecture on chemical science, " That elementary power was capable of being applied to propel carriages along the common roads with great political advantage, and that the floating knowledge of the day placed the subject within our reach." He soon afterwards constructed a little locomotive, which worked successfully with ammoniacal gas, and the results of his experiments were so satisfactory that he turned his attention to steam carriages. In 1825 he patented and made a very crude steam locomotive, impelled by legs similar to, but not nearly so well arranged as Brunton's and Gordon's arrangements, previously illustrated in these pages. In his experiments with this engine he ascended Windmill Hill, near Kilburn, and although many road locomotives had been constructed before Gurney's day, by which it was shown that propellers were altogether unnecessary (the wheels being found not only sufficient for propelling the carriages on level roads, but also for ascending hills), Gurney refused to believe in the tractive power of wheels, and provided his next carriage with levers to assist in propelling the carriage when the driving wheels slipped. A trip by steam between London and Edgeware, about nine miles from the carriage factory, was effected by this arrangement. In 1826 he constructed a coach about 20ft. long, which would accommodate six inside and fifteen outside passengers ; the driving wheels were 5ft., diam., and the leading wheels 3ft. 9in. diam. ; and two propellers were used, which could be put in motion when the carriage was climbing hills. Gurney's patent boiler was used for supplying steam to the 12 h.p. engine. The total weight of the carriage was about a ton and a half. This locomotive went up Highgate Hill to Edgeware, and also to Stanmore, and went up both Stanmore Hill and Brockley Hill. In ascending these hills the driving wheels did not slip, so that the legs were not needed. After these experiments the propellers were removed. In 1828 Gurney built a steam carriage, in which a number of patented improvements were embodied. Fig. 36 shows this steam carriage "which was propelled by the adhesion of one wheel, though means were provided for driving with both if required." He made many trips with this carriage, running to Barnet, and experimenting for eighteen months

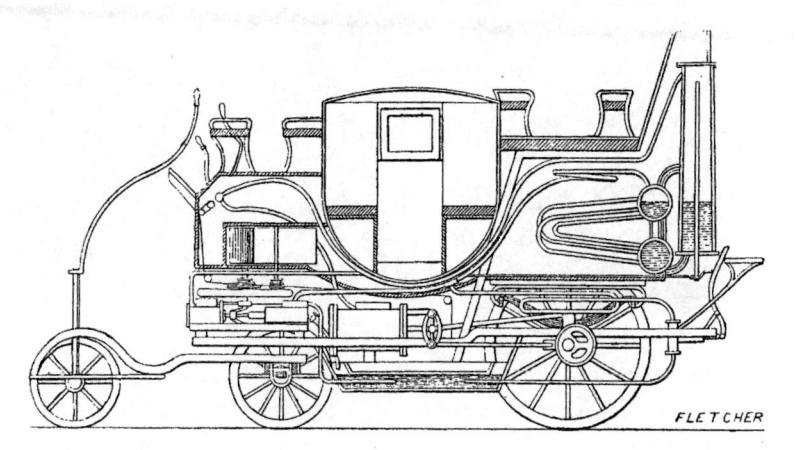

Fig. 36.

in the neighbourhood of London. It is reputed that he went to Bath and back, but one authority says that the coach broke in going, and was hauled into Bath by horses. Some repairs were effected in Bath, and during the return journey the carriage ran from Melksham to Cranford Bridge, a distance of eighty-four miles, in ten hours, including stoppages. This carriage, containing so many improvements, has been illustrated in numerous works on the steam engine, but we know that it was the least successful of all Gurney's attempts.* Fig. 37 shows the external appearance of the carriage by permission of the proprietors of *The Graphic*. In 1831 Sir Charles Dance commenced to run steam

Fig. 37.

carriages of Gurney's make between Gloucester and Cheltenham. These carriages ran for four months four times a-day, and during this period 3000 passengers were carried nearly 4000 miles. They performed the distance (nine miles) in fifty-five minutes on an average, and frequently did it in forty-five. "There were sometimes delays," owing to leaky boiler tubes, which prolonged the time, but no accident happened to any person. These carriages or "steam drags" are shewn in *Gordon's Elemental Locomotion* (1832), and are sensibly built engines. As will be seen from the illustrations,

* Maceroni says; "At a sale in 1834 a couple of those celebrated steam carriages, all but new (with a separate engine to work the pumps and blower), were sold for a mere trifle."

THE PERIOD OF SUCCESSFUL APPLICATION. 101

Fig. 38 gives a side elevation, and Fig. 39 a plan of the steam drag, the locomotive was distinct from the carriages in-

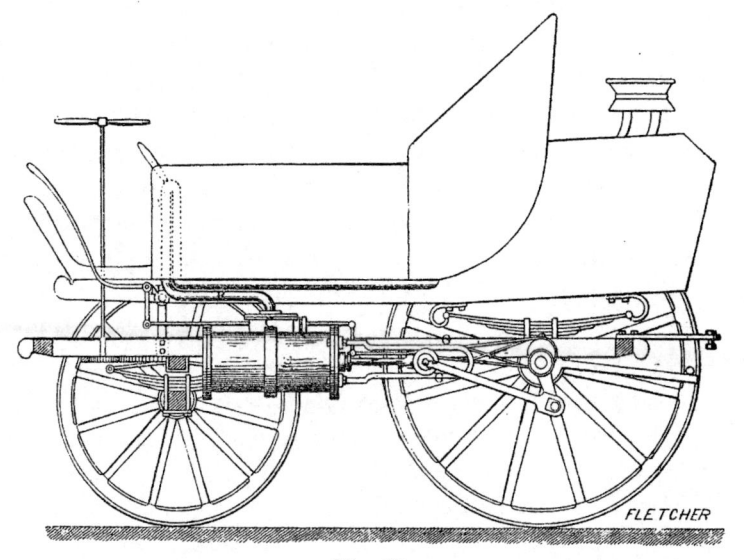

Fig. 38.

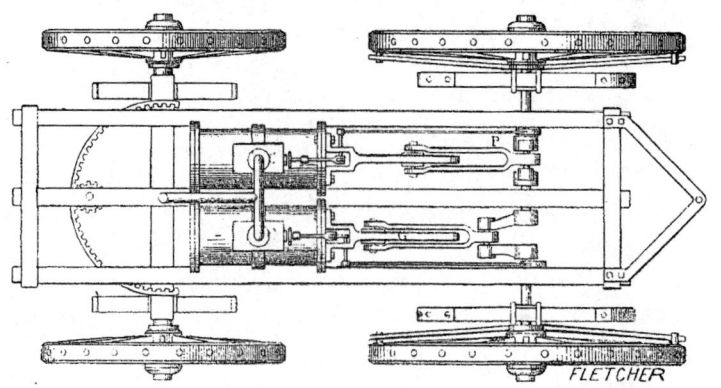

Fig. 39.

tended for passengers; the double cylinder engine was placed under the body of the "drag," and worked directly on to the cranked hind axle. The slide valves were actuated by eccentrics, and the steam pressure employed was 70lb. per square inch. Sir Charles Dance says: "Obstacles are always thrown in the way of a new invention, particularly if it is likely to produce important results, by those who expect their interests will be affected by its success." Thus it was that country gentlemen, trustees of roads, farmers, coach proprietors, coachmen, post boys, &c., opposed Gurney's steam carriages, and on the 22nd June, 1831, large heaps of stones were laid across the road, four miles from Gloucester, about 18in. deep; whilst struggling over this obstruction the axle was broken. Prohibitory turnpike rates ultimately turned these carriages off the road. In May, 1831, Mr. Ward made several trips in and about Glasgow with one of Gurney's carriages, but these trips were not so satisfactory as the owners could have wished. During a trip made with this steam carriage between Glasgow and Paisley the boiler burst, injuring some of the spectators. At the petition of Gurney, the House of Commons appointed a Select Committee to investigate the subject of steam locomotion on common roads. Many witnesses were examined during the three months that the committee sat. The evidence adduced was in favour of steam carriages.

About the year 1832 Sir Goldsworthy Gurney had ceased to build steam carriages at his own expense; he had just completed two "perfect" steam drags with other people's money, and refused to run them unless more money could be raised to build a third. He was very busy at this time attempting to float a company to introduce and run his improved carriages. One project was mooted of forming a joint concern of all the steam carriage proprietors, viz., Hancock, Maceroni, Redmund, and others, but this idea failed also. Gurney then petitioned the House of Commons to sanction a grant for the expenses he had incurred in attempting to introduce steam carriages, and to repay the heavy losses he had sustained through these numerous experimental trips. Sir George Cayley supported the prayer of the petition in Parliament. " He considered Mr. Gurney a very ill-used man. On the faith of our patent laws, Mr. Gurney had given up, to pursue this noble invention, a lucrative profession. Five years toil, and an expense of £30,000, brought it into practical use ; yet he is now deprived of receiving any remuneration—not from any want of success in his experiments, not from any failure in his carriage, but by Act of Parliament : by our act and deed has he been wronged, and by our act and deed ought he to be, and I trust will be, redressed." A select committee of the House of Commons recommended a grant of £16,000 and the repeal of the injurious Turnpike Acts. The Chancellor of the Exchequer refused the grant. The Steam Carriage Bill was twice referred to a select committee, had passed the House of Commons but was thrown out by the Lords. Mr. Gurney says : " When the repeal did not pass, I sold all my materials for manufacturing, and gave up my factory, feeling that injury had been done to me."

After Mr. Gurney quitted the steam carriage business he turned his attention to many other useful subjects—the production of the Bude light, the ventilation of mines, &c.—and in most of them he succeeded. The Bude light was tried for the first time in street illumination on the 10th January, 1842, at the crossing in Pall Mall at the bottom of Waterloo place. It is said to have illuminated the whole of the open space, in which stands the Athenæum Club, very powerfully, and to have caused the gas lamps to look as dim as the oil lamps did when gas was introduced. Gurney likewise devised the well-known stove which goes by his name. By these inventions his name was kept prominently before the public, and we are enabled to trace his career down to the time of his death, which occurred in March, 1875, at his residence in Cornwall.*

DANCE.—Sir Charles Dance, after removing Gurney's steam coaches from the Cheltenham and Gloucester road, came to London, and caused one of the carriages to be repaired and rendered more powerful by Messrs. Maudsley and Field. A patent was also taken out for a boiler in their joint names. In September, 1833, short experimental trips were made with the improved steam drag and omnibus attached, which travelled at sixteen miles per hour. A few days afterwards Sir Charles Dance left London for Brighton with the omnibus in tow, loaded with fifteen passengers, and covered the distance of fifty-two miles in five and a half hours, the return journey being performed in less than five hours. About the middle of October the steam drag and omnibus, says Young, "were put upon the road between Wellington Street, Waterloo Bridge, and Greenwich, where it continued to run for a fortnight, with a view of showing the public in London what could be done in this direction. The proprietor had no intention of making it a permanent mode of conveyance, and therefore kept the company as select as he could by charging half a crown for tickets each way."

CHURCH.—Dr. Church, of Birmingham, introduced a number of novelties pertaining to steam locomotion on common roads in 1832 and 1835. His patent specifications were very extensive and elaborate; the improvements consisted first, in the construction of the framing of the carriage which supports the bodies of the vehicle, and encloses the

* *The Engineer*, 12th March, 1875.

machinery by which it is impelled, and the manner of connecting the ribs for the purpose of giving great strength from a comparatively small weight of material; secondly, in the peculiar construction of the boilers, furnaces, and flues, by which a sufficient quantity of water may be converted into steam; and thirdly, in the construction of the running wheels of the coach, designed particularly to prevent concussions as they pass over inequalities on the surface of the road. The wheels were made with elastic rims that bent into "flatted curves" as they came in contact with the ground, thereby preventing the wheels from sinking or sliding round. The frame work of the vehicle was made of iron, and the boiler consisted of a series of vertical tubes, into each of which was introduced a pipe which was secured at the bottom of the boiler tube; the interior pipe constituted the flue, which first passed in through a boiler tube, and was then bent like a syphon, and passed down another until it reached as low or lower than the bottom of the fire place, whence it passed off into a general flue in communication with an exhausting apparatus. The wheels were elastic, the tyres being made of several successive layers of broad wood hoops, covered with a thin iron tyre. These elastic tyres were connected to the naves by spring spokes, very similar in construction to some modern traction engine wheels. Two oscillating steam cylinders were suspended on the steam and exhaust pipes over the crank shaft. The crank shaft and driving axles were connected together by chain wheels and chains, the wheels being made of different sizes for varying the speed by means of clutch boxes and levers conveniently arranged. The London and Birmingham Steam Carriage Company was formed in Birmingham with a large capital for the purpose of introducing Church's steam carriage. Large sums of money were advanced by the Company to Church, in the confident conviction that he would bring out a practical steam carriage. After various attempts, Dr. Church at length produced the steam coach shown by the illustration*, Fig. 40, the multitudinous combinations and the workmanship of which, says a competitor, were very scientific and beautiful. In 1835, this engine was brought out for public trial, and started from the factory with forty passengers, and proceeded

Fig. 40.

at a rapid rate for a considerable distance, but in turning round, the hind part struck the footpath and damaged some detail connected with the boiler; it was deemed unsafe to work it further, so the engine was hauled back to the works.

A few days afterwards the *Birmingham Gazette* said "The attention of many persons was attracted to Church's beautiful engine running on the Coventry Road, six miles out and six

* By permission of the Proprietors of *The Graphic*.

miles home, over heavy roads;" three miles were run on the level road at the rate of 15 miles an hour. Dr. Church, like most of his contemporaries, was not satisfied with a moderate speed, and consequently his carriage failed because it was impossible that such a complicated machine as he built would run at the rate of 15 to 20 miles on any common road without constantly breaking down.

A few months after the trip recorded above, the London and Birmingham Steam Carriage Company were advertising that all the difficulties of running steam carriages upon gravel roads were now overcome, and would be done to great profit to those engaged in it. It was wisely suggested that instead of puffing and advertising, the company should put a carriage on the road at once for passenger traffic between Birmingham and London; but this scheme was never practically accomplished; the carriages were constantly brought out, and as constantly failed. Previous to the opening of the London and Birmingham Railway, Dr. Church built the "Eclipse," the first four-wheel tank railway engine ever made. It was used as a ballasting engine on the above railway during its construction in 1838. He does not appear to return to the subject of steam on common roads. The only record we can find concerning Church after this date is that he sailed for America in 1861.

YATES AND SMITH.—On the 1st of July, 1834, a steam carriage, on a new principle, invented and made by Messrs. Yates and Smith, started from their factory, Colchester Street, Whitechapel, London, on its first trial. It ran up Whitechapel Lane, along High Street, and returned down Red Lion street and Leman Street to the factory, at the rate of 10 or 12 miles an hour. The exhaust pipe joint broke when running over the rough paving. When repaired the makers took the carriage for a trip on the Brighton Road. The engines were of the vibrating type, working in horizontal framing. The coach was similar to an ordinary stage coach.

FIELD.—After Sir Charles Dance had removed his steam coaches from the Gloucester and Cheltenham Road, he took one of them to Lambeth to be repaired by Messrs. Maudsley, Sons, and Field, the well-known marine engineers. The result was a patent taken out in the joint names of Sir Charles Dance and Mr. Joshua Field, for an improved boiler. Many journeys were made with this improved carriage, Mr. Field often accompanying Sir Charles Dance to Marlborough, Brighton, and other places. Mr. Field constructed a steam carriage which made an excursion in July, 1835, the passengers consisting of baronets and a select party. In the course of its journey, it went up Denmark Hill, and did the distance, nine miles, in forty-four minutes. It also ran several times to Reading and back, at the rate of 12 miles an hour. Mr. Cubitt, C.E., was one of the party who subscribed towards the building of Mr. Field's steam carriage, and he says that it was a success mechanically, but was far from being economical.

Mr. Field was one of the six gentlemen who founded the Institution of Civil Engineers. He died at the age of 77 years, in August 1863.

MILLICHAP.—In reply to a correspondent's letter to one of the engineering papers in 1837, G. Millichap, of Birmingham, says:.." If your correspondent will take the trouble to call at my house I shall be happy to show him a locomotive carriage in a state of great forwardness, intended decidedly for common roads."

This engine was on an entirely new principle, but we have no record of a public trial, so we are led to infer that the carriage was not a practical success.

HANCOCK.—Walter Hancock, of Stratford, London, was the most successful of all the steam carriage schemers, and to give anything like a detailed account of the many carriages built by him would occupy more time and space than is available; we must, therefore, be content to give a brief outline of his career during the twelve years that he devoted to the subject. In 1824, Hancock invented a novel form of engine, which he thought was very suitable for steam carriages, because of its simplicity, lightness, and comparative cheapness. A regular reciprocating motion was obtained by the alternate filling and discharging of india-rubber receivers, and this motion was given a rotary form by a crank in the usual manner. A 4 h.p. engine of this design was employed at the inventor's manufactory at Stratford, and worked well. Hancock also made a model steam carriage, and afterwards one on a larger scale; but only to discover, after many trials, that his engine was practically useless for steam carriage purposes. In 1827, he invented his well known boiler, which tended more than anything else to make his steam carriages in after years so successful. This boiler consisted of a number of flat chambers 2in. wide, ranged side by side, with about $\frac{3}{4}$in. space between each chamber. The sides of these chambers were covered with bosses, arranged so that the bosses of one chamber touched the bosses of the next chamber, thus forming abutments, and at the same time increasing the heating surface; the whole boiler was braced together by strong bolts and stays. The steam pressure used was 100lb. per square inch. The boilers were 2ft. square, and 3ft. high. Being satisfied with his boiler, Hancock next determined to build a steam carriage, using an engine of the ordinary construction. The carriage is shown in Fig. 41; it was supported on three wheels, and was intended to carry four passengers. The propelling force was obtained from a pair of oscillating cylinders working the double cranked axle of the leading wheel, which was arranged to swivel for steering. This carriage was far from being satisfactory, and was subjected to numerous alterations; but in spite of its defects it ran many hundreds of miles in experimental trips, from the manufactory at Stratford, "sometimes to Epping Forest, at other times to Paddington, and frequently to Whitechapel.

Fig. 41.

On one occasion it ran to Hounslow, and on another to Croydon. In every instance it accomplished the task assigned to it, and returned to Stratford on the same day on which it set out. Subsequently this carriage went from Stratford, through Pentonville, to Turnham Green, over Hammersmith Bridge, and thence to Fulham. In that neighbourhood it remained several days, and made a number of excursions in different directions, for the gratification of some of Hancock's friends, and others who had expressed a desire to witness its performance."* In 1831 Hancock commenced running his steam coach "The Infant" regularly

* Hancock's "Narrative of Twelve Years' Experiments," 1838.

THE PERIOD OF SUCCESSFUL APPLICATION. 111

for hire between Stratford and London, see Fig. 42. Before it was placed on the station it was tried in every possible way. The possibility of a steam carriage ascending steep hills had been doubted by many, and to remove if possible all scepticism on the subject, Hancock fixed a day for taking his carriage up Pentonville Hill, which has a rise of one in eighteen to twenty, and invited a numerous party to witness the experiments. He says : " A severe frost succeeding a shower of sleet had completely glazed the road, so that horses could scarcely keep

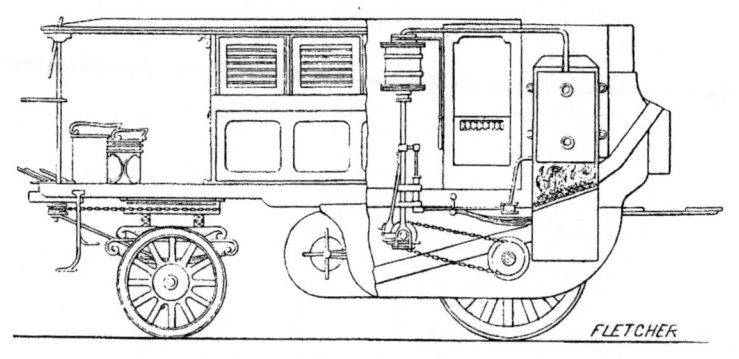

Fig 42

their footing. The trial was made therefore under the most unfavourable circumstances possible ; so much so, that confident as the writer felt in the powers of his engine, his heart inclined to fail him. The carriage, however, did its duty nobly. Without the aid of propellers or any other such appendages (then thought necessary on a level road) the hill was ascended at considerable speed and the summit successfully attained, while his competitors with their horses were yet but a little way from the bottom of the hill." Hancock's improved carriage was brought out during the latter part of 1832, in which the oscillating cylinders were abandoned, fixed cylinders taking their place. This carriage took a trip to Brighton and back, accompanied by Gordon and other scientific gentlemen, eleven in all. It ran at the rate of nine to fifteen miles an hour on the level road, and six miles an hour when ascending hills. In 1833, the "Autopsy" and the "Enterprise" were built. The "Autopsy" ran for hire between Finsbury Square and Pentonville, and continued to do so daily, without accident or intermission, for nearly four weeks. From August to November of 1834, Hancock ran the "Era" and the "Autopsy" for hire daily between the City, Moorgate, and Paddington, and during this period he carried nearly four thousand passengers, often running at a speed of twelve miles an hour.

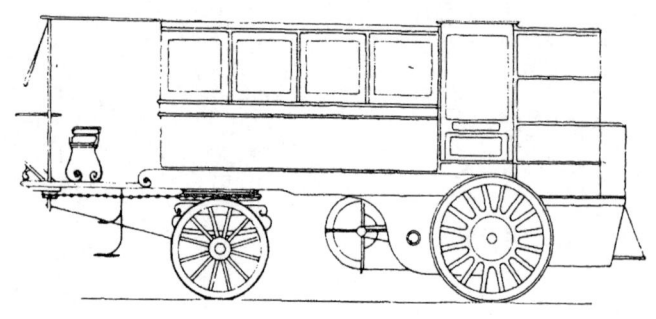

Fig. 43.

Hancock was the only steam carriage proprietor who had ventured to run a locomotive along the crowded streets of London at the busiest periods of the day. These hard roads were a severe test for the wheels and the gearing. The motion of these carriages was easy; they made no noise, and produced no smoke, and did not frighten horses. The "Era" was eventually shipped to Dublin, where it arrived safely in 1835. It ran through the principal streets and most crowded thoroughfares in Dublin with the most perfect success. On one occasion it ran three times round Stephen's Green at the rate of 18 miles an hour. In May, 1836, Hancock put the whole of his carriages on the Paddington road, and ran them daily without any intermission for upwards of five months, during which time they traversed 4,200 miles, made 525 trips from the city to Islington and back, 143 to Paddington and back, and 44 to Stratford and back : and the number of times he passed through the City came to more than two hundred. For five weeks he ran a carriage twice a day to the Bank. The following list of steam carriages built by Hancock, in the order of their construction, and the number of persons they were respectively calculated to accommodate, exclusive of the steersman, engineer, and fireman, will be of interest :— Experimental carriage, four outside ; "Infant" (trunnion engines), ten outside ; improved carriage (fixed engines) fourteen outside ; "Era," Greenwich, sixteen inside, two out ; "Enterprise," fourteen inside ; "Autopsy," nine inside, five out ; "Erin," eight inside, six out ; German drag, six outside, rest in carriages drawn ; "Automaton," twenty-two outside. The first time the "Automaton" was brought out upon the road it took a party to Romford and back at the rate of ten to twelve miles an hour. Young says : " On one occasion it performed (when put to the top of its speed, and loaded with twenty full grown persons) a mile on the Bow road at the rate of twenty-one miles an hour ! "

The Enterprise is represented by Fig. 43. It was built for the London and Paddington Steam Carriage Company, and commenced to run for hire under Mr. Hancock's personal superintendence between the city and Paddington for sixteen successive days, doing two or three journeys a day "to prove its capability of proceeding through crowded thoroughfares without inconvenience or liability to accident to the persons in the coach or others."

Mr. Hancock says, respecting these preliminary trips : " It is not intended to run this carriage more than about a week longer; partly because it was only intended as a demonstration of its efficiency, and partly because my own occupation will not admit of my personal attention to the steering, which I have hitherto performed myself, having no other person at present to whose guidance I could, with propriety, entrust it." The Enterprise ran from Cottage Lane, City Road to Paddington, and from Paddington to London Wall, and back to Cottage Lane, nine or ten miles, in less than an hour, exclusive of stoppages, performing the trips in an exceedingly satisfactory manner, and the carriage was more under the control of the driver than the best driven horse coach; it ascended Pentonville Hill with ease at six miles an hour. This steam omnibus, of course, was opposed by the drivers of horse vehicles. We quote the following from one of the daily papers: "In watching, as I have done, the early operations of the new steam coach the Enterprise, on the Paddington Road, I have been pained, though not surprised, to see the malignant efforts of some of the drivers of the horse conveyances to impede and baffle the course of the new competitor. They must be taught not to endanger the lives of the passengers, who have entrusted themselves to their guidance, by a wanton courting of collision with a vehicle so vastly more weighty, more strong, and more powerful than their own frail vehicles, and feeble, staggering beasts of draught. One of these infatuated men, to-day, crossed about the path of the steam coach, palpably with a mischievous design, which was only rendered abortive by the vigilance and prompt action of Mr. Hancock."

The London and Paddington Steam Carriage Company behaved very shabbily to Hancock. He had built the carriage to their order, had run it for several weeks over the course to test its powers at his own expense. Mr. Redmund, the engineer for the company, was satisfied with its perform-

THE PERIOD OF SUCCESSFUL APPLICATION.   115

ance after it had been running a week, and wrote a flattering report to the directors and shareholders respecting it. After the Enterprise had been delivered to the company, Mr. Hancock expected to receive directions to proceed with two more carriages upon the same plan, as per contract entered into by the company. " Various pretexts were, however, resorted to for delay, which subsequent proceedings proved were merely employed by the engineer as feints to conceal a design which for dishonesty has seldom been exceeded." During a delay which lasted for nearly six months, when a voluminous correspondence took place, Mr. Redmund had meanwhile taken the Enterprise to pieces, and was making a carriage of his own on the same lines. This we shall refer to in due course. Mr. Hancock at length caused the

Fig. 44.

Enterprise to be put together and returned, and thus ended the unprofitable and unpleasant business. We must briefly refer to the last and perhaps the best steam carriage built by Mr. Hancock in 1838, which is illustrated by Fig. 44. It will be seen from the illustration that the engine was of the vertical type, placed about the centre of the carriage. C is the cylinder ; the crank shaft works in bearings fixed to each side plate, on which is keyed a chain pulley, while a similar pulley of larger diameter was fitted to the driving axle ; a strong pitch chain communicated the motion from the crank shaft to the axle ; the up and down motion of the axle, which was hung on springs, not interfering with this method of driving. The boiler was situated at the back of the conveyance, the stoker's place being in the middle of the carriage. T represents a water tank and seat for two passengers. The steering arrangements are clearly shown by the illustration. The action of the little locomotive was most creditable to its builder.

In May, 1838, Mr. Hancock and two friends rode through the principal streets of the city in this steam carriage, caused it to run round the open space in front of the Guildhall, turn in any direction, run at any speed desired by its conductor, to the delight of a number of onlookers, after which Hancock threaded his way through the crowd of carts, omnibuses, cabs, and other vehicles in Cheapside, Leadenhall Street and other busy thoroughfares, stopped at the bank for a few minutes where Hancock alighted, leaving his friends in charge of the gig. One of the bank porters pompously ordered the gentlemen to " move on," but having had no experience with machinery they were placed in a dilemma, so they were obliged to confess their inability to comply with the order, to the great amusement of the bystanders. When its master arrived the locomotive moved off in good style and returned to Stratford. On the 22nd of June, Hyde Park presented an unusually gay appearance in consequence of a crowd of fashionable people being assembled to witness the trial of this little favourite steam carriage, which ran about among the splendid equipages for three or four hours without the slightest failure. Mr. Hancock guided it, caused it to turn in its own length, repeatedly stopping and starting it, then ran a distance at the rate of twelve miles an hour. The nobility who had met for the purpose of seeing it were delighted ; their horses, too, were not frightened, because it was noiseless, and it emitted no smoke or steam. In the early part of 1839, Hancock stated by advertisement that he had prepared his largest steam carriage, the Automaton, see Fig. 45, for traffic, and was ready to enter into engagements with responsible parties to run on any turnpike roads. An old inhabitant of Stratford, in June, 1839, said : " I have repeatedly noticed the performance of Hancock's carriages from the first of his experiments up to his present state of perfection in steam locomotion on common roads. A few days ago the Automaton ran from Stratford, through Ilford, and thence back through

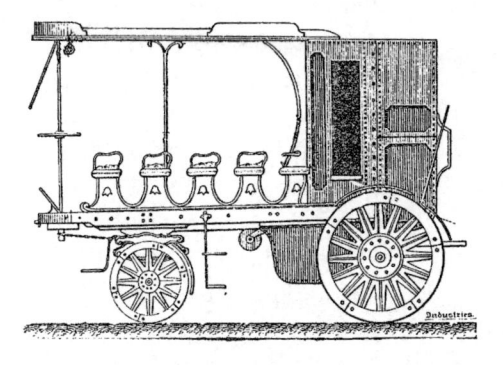

Fig. 45.

Stratford to the city, at fifteen miles an hour. Meeting the procession of the Lord Mayor and other city authorities going to hold a Court of Conservancy at the Swan Inn, Stratford, Hancock headed the procession to their destination, and in front of the house caused the carriage to perform a number of short trips and masterly evolutions, carrying at one time no less than thirty-two of the conservancy jury, quite to everyone's satisfaction."

The last trip we shall record is a novel one, performed as late as July 1840. A cricket club borrowed the Automaton of Mr Hancock to convey eleven of the Stratford Club and twenty-one visitors to the Forest. The run was a pleasant one, the carriage went properly with thirty-two people on board, although there were seats provided for only 22 passengers. The game was played, the Stratford team won, and returned to the factory gate at the rate of fifteen miles an hour. Numbers of persons who went to see the match occupied their time in viewing the conveyance that brought the players to the field, rather than the players themselves; but this personal neglect was good humouredly put up with by the club. Mr. Ogle says: Mr. Hancock, for want of support, was obliged to withdraw his carriages from the most difficult road in England, viz., the new road from the Bank to Paddington. Hancock says: "I entered upon that road, and continued running daily, solely with a view to demonstrate the practicability of so doing in the teeth of high authority to the contrary."

We now take leave of Mr. Hancock, the most successful locomotionist of those times, who, during sixteen years experience, built ten different carriages, each of which was creditably designed and made; and the later ones, as we have seen, were most successful. And the wonder is, how it happened that mere speculators, in several instances, who never made a steam carriage that would run the shortest distance without a breakdown, managed to float companies for the purpose of introducing their locomotives, which were not a success, while Hancock and Maceroni, the former a modest and retiring man, the latter rather boastful, both of whom built carriages far surpassing any of their contemporaries, were most unfortunate in their connection with the companies they were instrumental in forming for helping forward the steam locomotion movement. Sir Frederick Bramwell says: "It is quite certain that in respect of quietude of travelling, and in the way of not being an annoyance to others upon the road, Hancock's coaches of fifty years ago far exceeded anything of the present day. It may be asked why it was that if they were so meritorious in an engineering point of view, they did not continue to run? This is a difficult question to answer. Hancock always endeavoured to show that they paid, but it is believed that he was a better engineer and inventor than commercial man. Be this as it may however, it is unhappily the case that after many years of effort he gave up the endeavour."

REDMUND.—Mr. D. Redmund, City Road, London, while employed as engineer to the London and Paddington Steam Carriage Company, acted very dishonourably in secretly taking Hancock's new carriage, the Enterprise, to pieces, so that he could take dimensions and copy the design of the parts, to be embodied in a carriage he was making. He had previously patented a steam boiler, consisting of a series of vertical parallel chambers, an imitation of Hancock's patent. The manner of driving, the position of the engine, and even the external appearance, resembled the Enterprise closely. The driving wheels were of ornamental design, the subject of a separate patent, the cast iron spokes were of hollow section. They appeared to be the only part of the carriage not stolen from others, for Hebert says, "The steering arrangement was like Ackerman's patent of 1816." Before Redmund had finished his carriage, or had made any experiments, he boastfully advertised that he was willing to furnish locomotives to run on common roads at any required speed. When his steam carriage was announced as being ready for trial, the editor of the *Mechanics' Magazine* said, "We shall soon be able to judge whether he was justified or not in his confidence displayed in the advertisement." Redmund intimated that the private trials of his carriage were satisfactory, but publicity is the only test in such matters. But nothing was heard of Redmund's performances in public with his steam carriage, which he had named the Alpha. It was suggested that it might prove the Omega of his efforts in the steam engine line, and such turned out to be the case. The Alpha was a complete failure, and Redmund's project met with the fate it deserved. Fig. 46 shews the external appearance of Redmund's steam carriage.

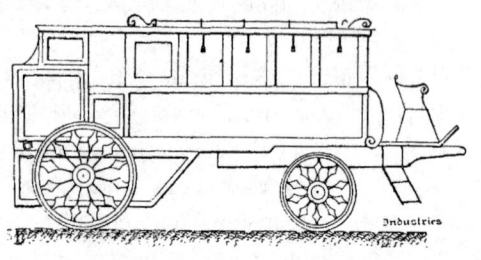

Fig. 46.

MACERONI.—Colonel Maceroni was one of Hancock's successful contemporaries, and among steam carriage promoters he occupies a very prominent position. Maceroni's father was an Italian merchant, residing in England, and for several years he occupied a quasi country house in the suburbs of Manchester, where in 1788 Francis Maceroni was born. We cannot afford the space to give any particulars of his early days or his many adventures. In 1814 he was living in Italy, and became aide-de-camp to the King of Naples. From 1825 to 1828 we find Maceroni helping Gurney, in London, to overcome the monetary difficulties associated with his early and somewhat faulty steam carriages ; but, feeling convinced that Gurney's efforts would never succeed, in 1829 Maceroni abandoned the steam carriage business and went to Constantinople. Having returned to England in 1831, "Mr. J. Squire came to me," says Maceroni,* " and informed me that

* "Memoirs of the Life and Adventures of Col. Maceroni," 1838.

he had built a steam carriage, which performed very well, and asked me to join him in the undertaking. Finding the little carriage much superior to any that Gurney had made, but unfortunately fitted with a somewhat defective boiler, I undertook to join in the construction of another on my plan, for which a valid patent could be obtained, but I was without much money, having, through the 'fortunes of war,' returned from Turkey with even less than I went out with. However, I mentioned my dilemma to a gentleman, the like of whom there are too few in this world, who provided me with the funds for taking convenient premises, purchasing lathes, tools, and establishing a factory on the Paddington Wharf. I placed Mr. Squire in the house on the works as foreman, and we set to work on an enlarged scale." In 1833 Colonel Maceroni and Mr. Squire jointly patented an efficient multitubular boiler, which was composed of eighty-one upright cylindrical tubes, disposed in nine rows, in the middle of which was the fire place. The tubes were all connected by horizontal tubes at the bottom and the top, the lower being a water communication, and the upper a steam communication. To prevent the formation of clinkers, and to preserve the fire bars from being rapidly burnt out, they were formed of hollow tubes, through which water circulated to the upright tubes of the boiler. The steam was conducted from the top of the tubes to a steam dome, from which the engine was supplied. The flame and heated matters were diffused around the whole series of tubes, and produced rapid generation of steam. The working pressure was 150lb. After completing this admirable boiler, Colonel Maceroni constructed a steam carriage, of which Fig. 47 is a representation. Mr. Gordon describes the carriage as "a fine specimen of indomitable perseverance," and he states that it not uncommonly travelled at the rate of from eighteen to twenty miles an hour. The engines were placed horizontally underneath the carriage body, the boiler was arranged at the back, and a fan was used to urge the combustion of the fuel, the supply of which was regulated by the engineman, who had a special seat behind. The passengers were placed in the open carriage body, and their seats were formed upon the tops of the water tanks. There were two cylinders 7½in. in diameter, the stroke being 15¾in. The diameter of the steam pipe was 2¼in., and that of the exhaust pipe was 2¾in. Colonel Maceroni's first steam carriage attracted much attention. From the commencement of his trials, he invited public investigation and publicity, and at a later date he writes: " My workshop doors, at 19, Wharf, Paddington, were open to every visitor ; and even from our very first experimental movements, I invited the editors of newspapers, engineers, and other authoritative and scientific persons to inspect our progress, and ride on the carriage any day, every day, and as often as would suit their pleasure or convenience." He asked the representatives of the Press to note the mile stones with watch in hand, and state the facts and make such observations as they thought proper ; and from these voluminous reports we will mention a few particulars of these interesting trials. " On October 4th, 1833, Colonel Maceroni and Mr. Squire, accompanied by eight other persons, took a trip in their new patent steam carriage, from Paddington Green to Edgware. The average speed was sixteen miles an hour. The return journey, a distance of 7½ miles, was performed in a little over thirty minutes. The carriage has gone to Windsor in two hours." " It appears to be the simplest and most compact steam carriage that has yet been tried in public."* " It is capable of carrying twelve or fourteen per-

* The *True Sun*, 5th October, 1833.

THE PERIOD OF SUCCESSFUL APPLICATION.

sons fourteen miles an hour, with perfect safety, on a turnpike road. It has been guided with ease and perfect security through Fleet Street and Cheapside at the most crowded time of the day." We also quote the following particulars, which were published in the *Morning Chronicle*, 14th October, 1833: "This steam carriage has plied daily for some weeks between Paddington and Edgware, without meeting with any accident. Since it was started, it has travelled a distance of upwards of seventeen hundred miles; yet, in the whole of that time, it has not needed any repairs." Colonel Maceroni once took a trip to Harrow-on-the-Hill, the distance of nine miles being completed in fifty-eight minutes. The hill was ascended with ease at the rate of seven miles an hour, and during no part of the journey was the full power of steam put on. For several weeks in the early part of 1834 the carriage was running daily from Oxford Street to Edgware. Afterwards several trips were made to Uxbridge, when the roads were in the worst possible condition, and nevertheless the journey from the Regent's Circus, Oxford Street, to Uxbridge, a distance of sixteen miles, was often performed in a little over an hour.

The following account of a trip to Watford appeared in "Turner's Annual Tour," in 1834: "Drawn out of a hut on Bushy Heath by the appearance of an unusual commotion amongst the inhabitants of the village, we saw a steam coach which stopped there. The apparition of a vehicle of this kind, in such a place, was unaccountable. Bushy Heath forms the plateau of a mountain, which is the highest point of land in Middlesex, and, although so far inland, serves as a landmark for vessels at sea. The access to it, from the London side, is by a difficult and steep road. Being accosted by Colonel Maceroni, in whom we were glad to recognise an old acquaintance, he informed us that the journey had been performed with ease, adding that it was his intention to proceed to Watford.

Now, if the road from Edgware to Bushy Heath was steep and difficult, the descent from Bushy Heath to Watford was much worse. We told our friend that he might go by steam to Watford, but that we were quite certain that he would not return by the same means of locomotion. Nevertheless, at his pressing instance, we consented to hazard our own person in the adventure. We set off, amidst the cheers of the villagers. The motion was so steady that we could have read with ease, and the noise was no worse than that produced by a common vehicle. On arriving at the summit of Clay Hill, the local and inexperienced attendant neglected to clog the wheel until it became impossible. We went thundering down the hill at the rate of thirty miles an hour. Mr. Squire was

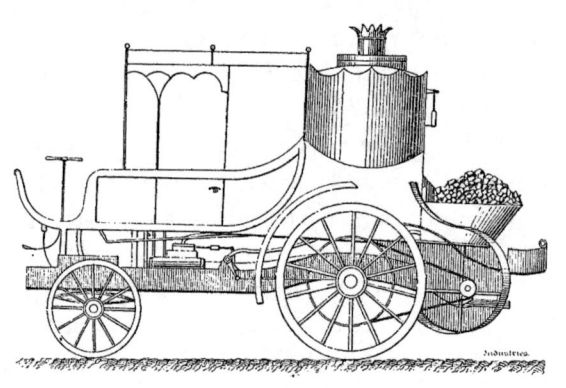

Fig. 47.

steersman, and never lost his presence of mind. It may be conceived what amazement a thing of this kind, flashing through the village of Bushy, occasioned among the inhabitants. The people seemed petrified on seeing a carriage without horses. In the busy and populous town of Watford the sensation was similar — the men gazed in speechless wonder; the women clapped their hands. We turned round at the end of the street in magnificent style, and ascended

Clay Hill at the same rate as the stage coaches drawn by five horses, and at length regained our starting place."

Maceroni had made two steam carriages, the first was intended to carry eleven persons, see fig. 47, and the excellent boiler worked at 150lbs. pressure per square inch. The second carriage was larger, and designed for carrying sixteen persons and ample room had been provided in this carriage for passengers' luggage.

In 1834 Maceroni and Squire dissolved partnership in the steam carriage business, after which Maceroni became very short of money, and allowed Asda, an Italian Jew, to take both the carriages to the Continent, he having promised to pay Maceroni £1,500 for a share in the patents taken out in France and Belgium. Asda solemnly stipulated that one of the carriages should be returned to England in six weeks. One carriage was running well in Brussels, long reports having appeared in the Belgian journals, stating that the performances justified all the expectations which had been formed respecting it.

Another carriage was doing equally well in Paris. The following account appeared in the *Journal de Paris* in February, 1835: "The steam carriage brought to perfection in England by Colonel Maceroni, ran along the Boulevards as far as the Rue Faubourg du Temple. It turned with the greatest facility, ran the whole length of the Boulevards back again, and along the Rue Royale, to the Place Louis XV. This carriage is very elegant, much lighter, and by no means so noisy as the one (Mr. Deitz's) we saw here some months ago, and it excited along its way the surprise and applause of the astonished spectators. All the hills on the paved Boulevard were ascended with astonishing rapidity. One of our colleagues was in this carriage the whole of its running above described, and he declares that there is not the least heat felt inside from the fire, and that conversation can be kept up so as to be heard at a much lower tone than in most ordinary carriages. The king took a ride in this carriage, and gave Asda a valuable present, who falsely styled himself the inventor, constructor, and sole proprietor of the carriages. A rich party of capitalists paid Asda £16,000 for the patent, of which poor Maceroni never received a shilling. All his tools and effects at the factory were taken by his creditors, and Maceroni was on the verge of starvation. In 1837 he attempted to form a company to construct and run steam carriages built under his directions, but the matter fell through, because he had no carriage to exhibit to those who had never seen either of those of which he had been so cruelly robbed. A year later another attempt was made to float a company, but without success. He made the following appeal, which, however, did not apparently meet with a practical response from the public: "If any party will provide the necessary funds to construct a couple of steam carriages according to my patent, I will engage, under any penalty or conditions that can reasonably be proposed, to run one of the same carriages from London to Birmingham and back to London within the time that it shall take any other steam carriage at present in existence simply to arrive at the same place. Both carriages to start at the same time. Ample guarantee will be given for the due and immediate construction of the carriages, which shall remain the property of the money provider, under fair and understood conditions."

However, in 1841, "The General Steam Carriage Company" had commenced to construct carriages in accordance with Maceroni's patents. Mr. Beale, of Greenwich, made the first carriage. The following report of the trip is given: "Having been accustomed to drive some of the best appointed fast coaches, I was invited to accompany a party of gentlemen on an experimental trip with Colonel Maceroni's steam locomotive. It started from Beale's Works, East Greenwich, and proceeded through Lewisham to Bromley, a distance of eight miles, performing the journey in half an hour ; we returned at the same rate. So confident was Beale in the performance of the engine that he determined to try Blackheath Hill, which was ascended in gallant style with a load of 17 passengers. We proceeded over Blackheath to the top of Shooter's Hill at the speed of 14 miles an hour, decending the hill, and reached the factory at a quick rate. Several shareholders were delighted." Maceroni had agreed to supply the carriages at £800 each to the Company. Mr. Beale's bill was £1,100 for his first locomotive, having charged over £200 for alterations and running some hundreds of miles on experimental trips. The Committee refused to pay this unreasonable amount, and Beale would not let the carriage go out any more. The quarrel was between the Board of Directors and the manufacturing engineer, but Colonel Maceroni was the greatest sufferer, because everything he had was seized by his creditors, his furniture, books, models, and he was now in great distress. Beale, in one of his letters, said :—" I believe Colonel Maceroni has done more than any other man in the kingdom towards steam locomotion on common roads, and if his scheme were properly supported it would succeed, and be of vast utility to the community." Towards the end of 1841 Maceroni offered for sale the patent rights of his steam boiler, which had been proved, by daily journeys on the most hilly roads. It had, during 18 consecutive months, propelled a steam carriage at the rate of 12 miles an hour, making little noise, and emitting no smoke. The patent had seven years to run, but in that time the advertisement went on to say, a great fortune might be made on common roads." Maceroni was a constant contributor to the *Mechanics' Magazine* from 1830 to 1840, but we fail to find a single line from his pen after his patent boiler advertisement had appeared, except a letter to the Editor in 1843, shewing that Squire's steam boiler, just then patented, was really an infringement of his patent boiler. Colonel Maceroni was most unfortunate in every transaction connected with steam locomotion.* He built two of the best road carriages ever made, and his labours have been entirely ignored by many writers on this subject.

A correspondent residing at Luton in 1840 said: "There has been so little written respecting steam road locomotion that I feared the matter was likely to fall through altogether. But I came across a gentleman whom I knew to have made experiments with a small steam carriage, who showed me two handsome and powerful carriages in his factory. One was complete, and had been out several times; the other was very nearly finished. The large one, with two cylinders, each 8 inches in diameter and 18 inches stroke, was intended to carry twenty passengers. The smaller one was built for conveying fifteen passengers. No expense had been spared to render them in every way a success." In addition to these engines, a large omnibus ready to attach to either of them had been constructed. We are unable to give the name of the maker of these carriages.

Colonel Maceroni about this time refers to some newly-designed road locomotives as follows: "There are three or four productions now being tried upon the Vauxhall Bridge and Finchley roads, but in mercy to the inventors I will not mention names, having seen their performances, which, like so many others, bring common road steam carriages into utter contempt.

* *The Cambridge Press* in 1839 said: "We have been long expecting Colonel Maceroni's steam carriage on the road from London to Cambridge, but we are weary of waiting." Maceroni (like Trevithick and many other steam coach builders) was continually hampered for want of funds.

DEITZ.—Mr. Deitz had run an engine about the streets of Paris previous to Maceroni's having been taken there, and in 1840 the engine was described in the reports of the Academy of Sciences and Academy of Industry in Paris, from which we quote a few particulars. The carriage had eight wheels, two of which were larger than the other six, and gave the impulsion. The six smaller wheels rose and fell according to the irregularity of the road, and at the same time assisted in bearing the weight of the carriages. The wheels were bound with wood tyres, having cork underneath. Deitz's locomotive was merely a drag; the passengers were drawn in a separate carriage. The engine was of 30 horse power, and travelled at a speed of ten miles an hour.

GIBBS AND APPLEGARTH.—In 1831 Mr. Gibbs gave evidence before the Committee of the House of Commons upon steam locomotion on common roads, in which he stated that he was building a steam carriage, and had travelled more than one hundred miles on Sir Charles Dance's coaches on the Gloucester and Cheltenham Road, in order to gain an experience that would be of service to him in his project, and to note the behaviour of the locomotives in passing over the rough roads and in mounting steep inclines. In the early part of 1832 Gibbs completed a steam drag. The chief characteristic of this locomotive, which was intended for drawing the passenger carriages behind it, was the boiler, with a spirally descending flue placed behind the driving wheels. In September, 1832, Gibbs and Applegarth patented a superior steam carriage with a tubular boiler and oscillating engine cylinders. Two spur pinions on the crankshaft of different diameters geared into two wheels on the main axle, either pair of which could be geared together and clutched to the driving axle for running the engine at two speeds. The power from the axle was transmitted to the driving wheels through friction bands, arranged on the bosses of the wheels, so that one or both wheels could be coupled to the axle. Friction driving bands similar to those introduced by Gibbs and Applegarth were used on some traction and steam ploughing engines until quite recently. Very little is said respecting the trials of the above engines; special mention is made by Gordon in reference to the boiler with the spiral downtake, which was said to be a good steamer.

WATTS.—About this time a locomotive was built by Mr. Watts, engineer, Norwich, for running from Norwich to Yarmouth, which answered, we are told, exceedingly well.

ROBERTS.—Mr. Roberts, of the celebrated firm of Sharp, Roberts, and Co., engineers, Manchester, constructed a road locomotive, which was subjected to a public trial in December, 1833, which, while it served to reveal a few constructive imperfections easily removed, tended to establish the soundness of the principle on which the carriage was constructed. The second trip took place in March, 1834. The carriage started from the works in Falkner Street, under the guidance of Mr. Roberts, with forty passengers. It proceeded about a mile and a half up Oxford Road, where, owing to the apprehension of a deficiency of water, a sudden turn was made, which was attended with some difficulty, as the road was narrow; it then proceeded back to the works, where it arrived without accident. The maximum speed on the level was nearly twenty miles an hour. Hills were mounted easily No doubt existed of the engine being speedily put in complete and effective condition for actual service. We regret to say that during another experimental trip in April, 1834, an accident occurred to this locomotive, which was reported to have been more serious than it actually was. Maceroni

THE PERIOD OF SUCCESSFUL APPLICATION. 131

grossly exaggerates the accident by stating that "the boiler burst in the streets of Manchester, severely wounded several persons, and set fire to an apothecary's shop."* It appears that the engine had travelled more than a mile from the works, when it was found that the pump on the engine did not work properly. The water in the boiler being dangerously low, the fire was quickly drawn; the boiler was filled with water to the proper level from a wayside pond, and the fire re-lighted. Mr. Roberts directed the carriage to be turned round, and it soon commenced its journey home, carrying from forty to fifty persons. It proceeded at a fair rate until it arrived near the corner of Rusholme Lane, where some of the boiler tubes gave way, and the steam escaping, blew the firebars down, and scattered the coke about in all directions, breaking several squares of glass in three shop windows near, slightly hurting a man and a boy who were hanging on behind the carriage. No one was seriously injured and none of the crowd of passengers were hurt. The carriage was removed to the works, drawn by four horses.

Several writers on steam road locomotion have accorded the credit of the invention of compensating gear to Mr. F. Hill, of Deptford, but it is our pleasure to be able to show that Mr. Roberts, the celebrated Manchester engineer, was the inventor of the compensating gear, which he used on the steam carriage we have just described. This compensating or differential gear is a device that superseded claw clutches, friction bands, ratchet wheels, and other complicated arrangements for obtaining the full power of both the driving wheels, and at the same time allowing for the engine to turn the sharpest corner without any difficulty. This compensating arrangement, introduced by Mr. Roberts nearly sixty years ago, is now universally adopted on modern traction engines.

* "Steam Power on Common Roads." Maceroni, 1835.

We give an illustration of this gear in a subsequent article.

Dr. Smiles, in his delightful book, "Industrial Biography," says that in 1839, Mr. Roberts invented an arrangement for communicating power to both driving wheels at all times, whether turning to the right or left.

Mr. Carrett used the gear on a little locomotive which he exhibited at the London Exhibition of 1862, and from an accurate description of the gear given by him, he admits that Roberts was the inventor. It is to be regretted that no particulars appear to have been given respecting the design of Roberts's interesting road locomotive. Emanating as it did from a famous engineering works, it doubtless far excelled in arrangement and workmanship the steam carriages produced by contemporary makers.

Mr. Roberts died in straitened circumstances in March, 1864, at the age of seventy-five years. One writer says Mr. Roberts "helped others in their difficulties, but forgot himself. Many have profited by his inventions without even acknowledging the obligations they owed to him. They have used his brains and copied his inventions, and the inventor is all but forgotten. It is lamentable to think that one of the most prolific and useful inventors of his time should in his old age have been left to fight with poverty." As is our usual method of treating our benefactors, we allowed Mr. Roberts to live in obscurity and die in want, and after his sufferings were ended a memorial was reared to his memory.

RUSSELL.—Mr. John Scott Russell (the well-known designer and builder of the Great Eastern) in early life took great interest in steam locomotion. He made a small steam carriage which ran about the neighbourhood of Greenock successfully. In later years when residing in

THE PERIOD OF SUCCESSFUL APPLICATION. 133

Edinburgh, he patented a steam locomotive intended for the conveyance of passengers on common roads. Six of these coaches were built under his patents and to his designs, by the Grove House Engine Works, Edinburgh, for the Steam Carriage Company of Scotland.

In April, 1834, this company established a line of steam coaches for the conveyance of passengers between Glasgow and Paisley, which plied hourly for many months with the greatest regularity and success. The distance between the two places was a little more than seven miles, and the trip

Fig. 48.

was run in 40 to 45 minutes. Mr. Russell's coaches were very popular with the travelling part of the community, and were repeatedly overcrowded, 30 to 40 persons finding places on a vehicle and its tender, constructed to carry six inside and twenty outside passengers. These carriages have been briefly referred to by two or three writers on this subject, but they have not been illustrated and described in any recent work on steam locomotion; in fact, these coaches have been practically omitted by previous writers, and in order to supply the missing link in the history of steam on roads, we devote a considerable amount of space to their description.

Mr. Scott Russell, as an experienced engineer, designed his coach with great care. Fig. 48 shows a side view of the carriage, while Fig. 49 gives an end view of the engine to an enlarged scale. The general appearance was far superior to many of its competitors, and we are told that " it

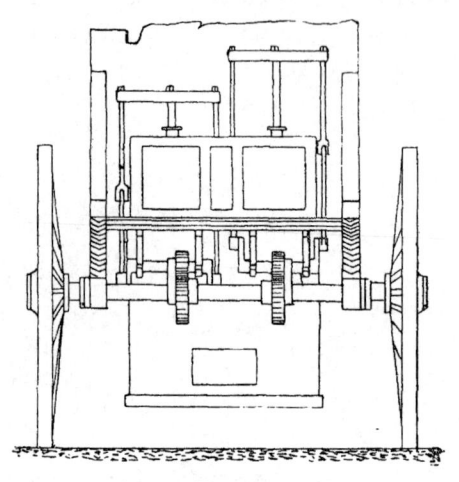

Fig. 49.

was fitted up in the style and with all the comfort and elegance of the most costly gentleman's travelling carriage." The boiler was of the multitubular type, with the furnace and the return tubes on the same level, and similar to a marine boiler. The improvements introduced by Russell consisted in constructing the boiler in such a manner that it should everywhere consist of opposite and parallel surfaces, or as nearly so as circumstances allowed, and connecting these surfaces together by means of stays of small diameter, placed at distances proportioned to their direct cohesive strength, and to the degree of pressure to be resisted; the plates were much thinner than usual, so that the heat was transmitted quicker; the copper plates were one-tenth of an inch thick. The stays were only one-quarter of an inch in diameter, there being thirteen hundred of them used. Mr. Russell said that the boiler was safe, because he thought the moment the pressure exceeds the maximum, the weakest of the stays will give way; and one rod giving way will instantly let out the whole of the water in the boiler, take off the pressure, extinguish the fire, and prevent all chances of explosion. But we regret to find that Russell's ideas respecting the safety of his boiler were not realised in practice, as we shall see presently. The whole weight of the carriage was supported on springs. The engine had two vertical cylinders, twelve inches diameter and twelve inches stroke. The piston rods worked through the top cylinder covers, and were connected by crossheads to two side connecting rods; the rods from each cylinder worked on to a separate crank shaft, as shown by Fig. 49. Each cylinder had four ports, which were alternately opened and closed by slide valves, actuated by eccentrics keyed on the crank shafts; one pair of these ports were for the admission of steam, and the other for the exhaust. On each crank axle was fixed a spur pinion gearing into a wheel on the main driving shaft—ratio, two to one; the crank shaft and the driving axle being coupled together by sun and planet straps, which kept the toothed wheels properly in gear. The engine was mounted upon laminated springs, so beautifully arranged that each spring in its flexure described, at a particular point, such a circle as was also described by the main axle in its motion round the crank shaft; thus any irregularities in the road in no way interfered with the proper working of the spur gearing. The exhaust steam was turned into the chimney to create a blast.

The water and coke were carried on a separate tender on two wheels, coupled to the rear of the engine; at different stations on the road spare tenders were kept in readiness, filled, and were quickly connected to the coach. This tender was mounted upon springs, and provided with seats back and front for passengers. India-rubber tubes conducted the water from the tank to the two brass feed pumps on the engine. Three persons were required to be in attendance—a steersman on the front seat, an engineer on the back seat outside above the engines, the stop valve and cocks being within his reach; he could also tell the height of the water in the boiler, and the amount of steam pressure. The stoker stood on the foot-plate in front of the boiler. These coaches were admirably worked out, and were said to be a "triumphant success" after they had run regularly for four months. Russell's coaches shared the same fate in Scotland that Sir Charles Dance's did in England. They had not been running many months before the road trustees at the Glasgow end, in order to cause an obstruction, put a thick coating of loose stones on the road, but the steam carriages ploughed through it. More men were then employed by the determined obstructionists to put another thick layer of stones on the top, so that the road was all but impassable. Ordinary road carriages were injured thereby, and heavy carts were obliged to desert the road, and go round by a different and much longer route. After the steam coaches had travelled over this accumulation of road material for some time one of the wheels broke, and the carriage was nearly overturned. The whole weight of the vehicle rested on the boiler, and caused it to burst, and five of the passengers were killed.

The Court of Session, in consequence of this accident, interdicted the whole set of carriages from running, for the time at least. The editor of the *Mechanics' Magazine* said this was a fine specimen of Caledonian wisdom! Why not clear the Clyde of steamers, because accidents happen with steamers as well as with carriages? The Steam Carriage Company brought an action for damages against the trustees of the turnpike road for having compelled them to give up the running of the carriages on the Glasgow and Paisley road by "wantonly, wrongfully, and maliciously accumulating masses of metal, stones, and rubbish on the said road, in order to create such annoyance and obstruction as might impede, overturn, or destroy the steam coaches belonging to the plaintiffs."

Russell's steam coaches were no longer used in Scotland, but two of them were sent by steamer to London, and were often engaged in running with passengers between London and Greenwich, or Kew Bridge. Several trips were made to Windsor. They were eventually offered for sale, and to show their powers they started every day at one o'clock from Hyde Park Corner to make a journey to Hammersmith. But they remained unsold, and we hear nothing further respecting them.

Mr. J. Scott Russell, however, was actively employed in shipbuilding, his name being a "household word" in everything pertaining to steam navigation. He was a Fellow of the Royal Society, vice-president of the Institute of Naval Architects, and a Member of Council of the Institute of Civil Engineers. A contemporary, in speaking of Mr. Russell's death, which occurred as recently as 1882, said, respecting the coaches we have illustrated: "The springs of his steam carriages, and the manner in which the machinery adapted itself to the irregularities of the road, were triumphs of engineering."

HILL.—Mr. F. Hill, of the Deptford Chemical Works, must be classed among the successful steam road locomotionists. We meet with him first in 1839 among the distinguished passengers who accompanied Mr. Hancock in one of his latest trips with the Automaton from London to Cambridge and back.

Mr. Hill was doubtless taking a lesson in steam carriage construction during the journey. In after years he devoted much time and attention to the subject, and was very successful in his carriage experiments. In 1840, Mr. Hill made repeated trips to Sevenoaks, Tunbridge Wells, &c., with satisfactory results. He also ran on the Brighton road, up steep hills, with the carriage fully loaded, at twelve miles an hour, and on the level at sixteen miles an hour. We find

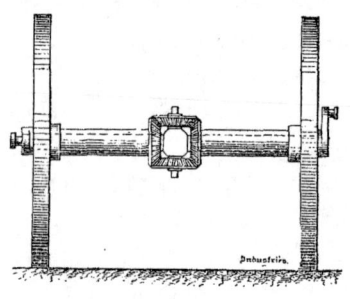

Fig. 50.

that Mr. Hill used the compensating gear among his steam carriage improvements, a device invented by Mr. Richard Roberts, of Manchester.

One of the difficulties attending the construction of steam carriages was the connection of the driving wheels with the machinery, so as to obtain the full adhesion of the wheels, and at the same time to allow facility in turning sharp corners. James, as we have noticed, fixed each driving wheel upon a separate axle. Hancock and others employed only three wheels in their carriages, and applied the power to the front wheel, which ran in advance of and between the tracks

THE PERIOD OF SUCCESSFUL APPLICATION. 139

of the other two. The most common plan, however, was to have only one wheel keyed to the driving axle, while the other was connected when required by a sliding clutch, which was anything but a convenient arrangement. But in the patent compensating gear used by Roberts, Hill and others, shown in Fig. 50, all these objectionable plans were obviated; and this simple and efficient gearing, introduced sixty years ago, is in regular use on every road engine built at the present time.* Upon examining the illustration it will be seen that so long as the driving wheels continue to run in a

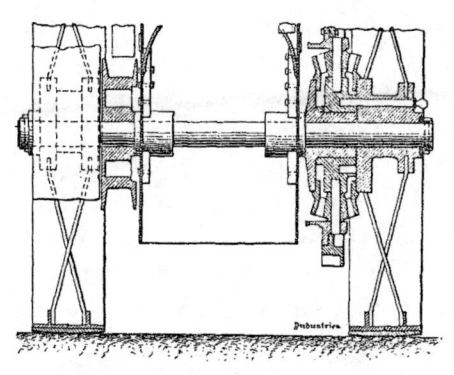

Fig. 51.

straight line, the tubes do not revolve upon the axle, but turn round with it, and carry round the wheels as if they were fixed to the axle; but when any deviation from the straight line takes place, the wheels, while advancing with the axle, revolve more or less upon the axle in contrary directions, so that the advance of the outer driving wheel exceeds that of the inner wheel by as much as the length of the outside curve exceeds that of the inner curve, and thus skidding is prevented by this ingenious arrangement. Fig. 51 shows the

* Mr. John McLaren, A.I.C.E., in a paper entitled, "Steam on Common Roads," read before the Institute of Civil Engineers in November, 1890, says: The compensating gear was first employed in White's dynamometer, and published in his "Century of Inventions," 68 years ago.

compensating gear as applied to a modern road locomotive; here one bevel wheel is bolted to the right hand side travelling wheel boss, while the other is keyed on the axle, and the left hand side driving wheel is likewise keyed to the axle. The full power of the engine is transmitted through the two bevel pinions, and both travelling wheels act as drivers, whether going in a straight line or not.

In August, 1841, the General Steam Carriage Company was formed for working Hill's patents. It was urged by the promoters that the demand for additional accommodation on some roads really existed, because it was desirable that road locomotion should counteract the exorbitant charges made by the gigantic railway monopoly for conveying goods short distances. The company state in their prospectus "that while they confidently believe the improved steam coach which they have engaged and propose to employ in the first instance to be the most perfect now known in England, they do not bind themselves to adhere to any particular invention, but will avail themselves of every discovery to promote steam coach conveyance." Instead of making short and showy trips on good suburban roads, Mr. Hill selected for his curriculum those roads which, from the peculiar difficulties they presented, were likely to point out every variety of provision that need be made, or circumstances that were to be guarded against. The Windsor, Brighton, Hastings, and similar roads, had been traversed with uniform success. Perhaps no more satisfactory performance could be cited of a common road steam carriage making a trip to Hastings and back, a distance of 128 miles, which was performed in one day, it being accomplished in half the time occupied by the stage coaches. The Editor of the *Mechanics' Magazine* said: "We accompanied Hill, about a year ago, in a short run up and down the hills about Blackheath, Bromley, and neighbourhood; and we had again the pleasure of accompanying him in a delightful trip, on the Hastings Road, as far as Tunbridge and back. The manner in which his carriage took all the hills, both in the ascent and the descent, proved how completely every difficulty on this head had been surmounted; Quarry Hill rises 1 in 13, River Hill, said by coachmen to be the worst hill in the country, rises 1 in 12." We illustrate Hill's steam carriage by figure 52. It will be seen from the engraving that the whole of the coach and machinery were erected upon a strong frame mounted upon substantial springs; the hinder part was occupied by the

Fig. 52.

boiler, furnace, and water tanks, with a place for the engineer and stoker. In front of these was a coach body capable of holding six people inside, three on the box, and the conductor in front. The front part of the carriage was suspended upon springs also, making the motion delightfully easy and agreeable. The carriage was propelled by a pair of 10in cylinders and pistons, lying horizontally beneath the carriage, which acted upon two 9in. cranks, which were coupled to the main axle through the compensating gear already referred to; the two 6ft. 6in. diameter driving wheels, had the full power of the engines passed through them, yet in case of any differential velocity required by either wheel when turning corners, the compensating bevel wheels revolved and thus allowed the engine to turn about any way. The weight of the boiler when empty was 23 cwt. When filled it held about sixty gallons of water, and one hundred gallons more were contained in the tanks which surrounded it. The quantity of water taken in at each of the stations (which were arranged as nearly as possible in eight-mile stages) was about eighty gallons. The total weight of the carriage, including water, coke, and twelve passengers, was less than four tons. When working on heavy and rough roads the steam pressure was seventy pounds per square inch, but on good roads sixty pounds was amply sufficient. The average travelling speed was sixteen miles an hour; on a level road the speed of twenty miles an hour has been realised. In long journeys, however, on public roads, the speed was regulated more by the casual obstructions arising from the ordinary traffic than by the power of steam. Mr. Hill's long experience proved to him that passengers could be conveyed by steam coaches at half the expense, and at double the speed of the stage coaches.

ANDERSON.—Sir James Anderson, Bart., of Buttevant Castle, Ireland, as far back as 1827 was engaged in the steam carriage enterprise in connection with Mr. James, of Holborn, the former, doubtless, finding the money to enable James to patent his inventions and carry out many experiments, until Anderson fell into pecuniary difficulties, which caused a dissolution of partnership, which was shortly followed by the discontinuance of James's experiments. But ten years afterwards we find the Baronet of Buttevant Castle engaged in steam carriage construction on a grand scale on his own account. In 1838 he says: "I have spent two apprenticeships to this undertaking, and have expended £30,000 on experiments."

Sir James Anderson first of all patented a boiler suitable for steam carriages, which we are told was a poor copy of Hancock's boiler. Maceroni says: "Sir James Anderson's plagiarism on Hancock's boiler will bring steam road travelling into contempt." While Hancock says; "The flat chambers, as arranged, cannot succeed, and really are an infringement of my patent;" but he says, "the boiler is sure to fail, so he need take no further trouble respecting it."

A joint-stock company was launched, having for its object the introduction of Sir James Anderson's steam carriages on common roads, termed the Steam Carriage and Wagon Company. The prospectus intimated that several steam drags were in course of construction in Dublin and in Manchester, which, when completed, were to convey goods and passengers at double the speed and at half the cost of horse carriages. Sir James Anderson says: "I produce and prove my steam drags before I am paid for them, and I keep them in repair; consequently, neither the public nor the company run any risk, the first steam carriage built for the company is nearly completed. It will speak for itself." His friends said he had failed in twenty-nine carriages to succeed in the thirtieth. In the *Mechanics' Magazine*, June, 1839, a Dublin correspondent writes: "I was fortunate enough to get a sight of Sir James Anderson's steam carriage, with which I was much pleased. It had just arrived from the country, and was destined for London in about three weeks. The engine weighs 10 tons, and will, I dare say, act very well. I shall have an opportunity of judging that, as the tender is at Cork. It has a sort of diligence, not joined, but to be attached to the tender, making in all three carriages. I talked a great deal about it to one of his principal men, who was most lavish in its praises, especially as regards the boiler." In July, 1839, it was announced by the papers that the vessel, having Sir James Anderson's steam carriage on board, sailed from Dublin for London, and was hourly expected, and upon arrival the carriage would be put together as quickly as possible, and submitted to such a trial as the directors of the Company shall direct. In August, the long expected carriage had at last arrived, and was undergoing the finishing touches of a London engineer, previous to its essay at locomotion on the roads of the metropolis. Meanwhile Hancock's "Automaton" was kept in readiness to compete with the long-vaunted powers of the machine of the Knight of Buttevant Castle. Colonel Maceroni was also preparing to take the road with his never-failing fountain of vapour. So we may judge that the result of the competition was awaited with interest, but the event does not appear to have taken place. In 1840, Mr. J. Rogers says: "Several steam carriages are being built at Manchester and Dublin, under Sir James Anderson's patents, and one has been completed at each place. At Manchester the steam drag had been frequently running between Cross Street and Altrincham, and the last run was made at the rate of twenty miles an hour, with four tons on the tender, in the presence of Mr Sharp, of the firm of Sharp, Roberts and Company, of Manchester, and others." The *Morning Herald* for 30th June, 1840, contained a long account of the doings of these steam drags, from which we gather the following particulars: An experimental trip of Sir James Anderson's steam drag for common roads took place on the Howth Road, Dublin. It ran about two hours, backing, and turning about in every direction—the object being chiefly to try the various parts in detail. It repeatedly turned the corners of the avenues at a speed of twelve miles an hour, the steam pressure required being only forty-six pounds per square inch. No smoke was seen, and little steam was observed. The whole machinery was ornamentally boxed in, so that none of the moving parts were exposed to view, and it was found that the horses did not shy at this carriage. The directors of the company were to assemble at Manchester, in order to witness a trial of the steam carriage constructed there, after which a meeting was to be called at Dublin for the purpose of forming a company in conjunction with one already established in England, for opening up a communication by means of these drags between the chief towns in Ireland, as soon as a few of the steam carriages were finished. It was proposed that the united company should in the first instance, in conjunction with the railway trains from London, run from Birmingham to Holyhead, the passengers to be thence conveyed to Dublin by steamer; from Dublin to Galway the steam drags were to be employed; and thence to New York per vessel touching at Halifax; thus making Ireland the stepping-stone between England, Nova Scotia, and the United States of America. It will be seen that Sir Jas. Anderson purposed great things with his steam carriages, but judging from the paucity of the literature dealing with his public trials, we fear that little practical running was accomplished.

Want of success, however, does not appear to have caused him to relax his energies, for in addition to the carriages we have referred to, many more schemes were proposed and patented, the latest bearing the date of 1858, showing that Sir James Anderson, Bart., devoted no less than thirty-one years of his life to the furtherance of steam locomotion on common roads.

L

SQUIRE.—In 1843, a tubular steam boiler intended for common road carriages was patented by Mr. Squire, whom it will be noted was ten years before in partnership with Colonel Maceroni. The steam boiler was the matured result of experiments in which the inventor had for several years been engaged, to introduce steam travelling on common roads. It is illustrated in the *Mechanics' Magazine*, vol. 39.

We have now arrived at a period in the history of steam on common roads where an immense gap occurs in the narrative. From 1840 to 1857 no new steam carriages were constructed, or, if any were made, no particulars of their design or record of their trials have been chronicled. A correspondent in the *Mechanics' Magazine*, in 1843, says: "Norrgber, of Sweden, a locksmith and an ingenious mechanic, made a steam carriage which ran between Copenhagen and Corsoer, carrying thirty passengers, the engine being of eight horse power." Mr. John Bourne, in his work entitled "Recent Improvements in the Steam Engine," refers to an arrangement proposed by himself in 1843, whereby the power of the engine of a common road locomotive might be communicated to the wheels without interfering with the free action of the springs; and he says: "Ever since 1847, when I first went to India, I have continued to urge the employment of suitable locomotives upon the great roads of that country."

It is true that agricultural locomotives were being developed during this period, but the passenger carriages appear to have been driven off the roads by the road commissioners, who levied heavy tolls upon pleasure steam carriages passing through the tollgates. A Leeds correspondent in a newspaper, in 1848, says: "The demand for such a toll will put a stop to any engineering project relating to road locomotion, even were the invention ever so plausible."

FISHER. — Mr. J. K. Fisher, of New York, in 1840, designed a small steam carriage similar in appearance to a railway locomotive, but he was advised by engineers not to pursue the subject further, as the English engines on much better roads had failed. In 1853 he built a small steam carriage, which had driving wheels 5ft. in diameter, and two steam cylinders 4in. in diameter, by 10in. stroke. The boiler was composed of a number of water tubes placed around the fire; this carriage attained a speed of fifteen miles an hour on good pavements, but was much too light and fragile for rough roads.

From 1859 to 1861 many trips were run with Mr. Fisher's steam carriages, some of them running twelve miles an hour without excessive wear. In his later engines he introduced several novelties. The first was the parallel connections between the crank shaft and the driving axle. The axle was held in place by radius rods, which were jointed at one end to the axle, at the other end to the crank-shaft carriages. The effect of this parallel connection was to prevent the connecting rods from causing the carriage to rock and pitch, as they would inevitably cause it to do if they acted on the guide bars at one end and on the wheel at the other —the guide bars being on springs, and the wheels on the road, and the connecting rods acting obliquely. It is understood that the bending of the springs by the action of the engines was one of the principle difficulties of the early steam carriages; and many of them were unable to have their engines suspended on easy springs, excepting those worked by driving chains like Hancock's.

Fig. 53 shows the steering gear used by Fisher from which it will be seen that a screw was placed across the front part of the carriage carrying a nut, to which the end of an elongated reverted pole was jointed, as shown. The screw was turned by bevel gearing, one wheel being keyed to the end of the screw, and the other keyed to the steerage rod, the opposite end of this rod having a lever placed within easy access of the footplate. The engines could be steered with facility, the screw worked well, and was less liable to be jerked out of position than the pinion formerly used for this purpose. For light steam carriages running on smooth roads, a simple tiller or guiding wheel was found sufficient, but on rough roads was very fatiguing to the hands. Mr. Fisher's carriages were driven by direct-acting engines, one cylinder on each side of the smoke-box, similarly arranged to an ordinary railway locomotive. His engines were a

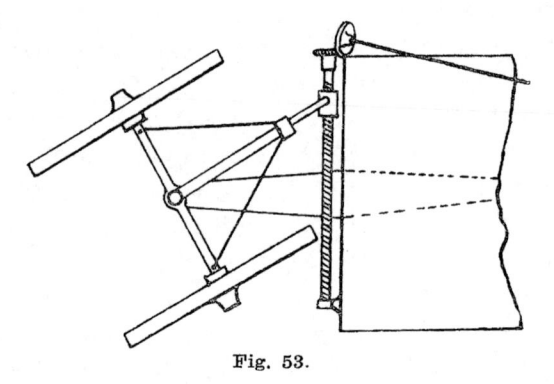

Fig. 53.

mechanical success, but from the public he met much discouragement. It is true he had no prohibitory tolls to pay, and none of the road trustees were base enough to copy the example set by the British authorities, who placed about eighteen inches of loose stones on two highways for the purpose of stopping Dance's carriages on the Gloucester road, and Russell's on the Glasgow road.

DUDGEON. — Mr. Richard Dudgeon, of New York, built a small locomotive for common roads in 1857. It

THE PERIOD OF SUCCESSFUL APPLICATION. 149

had two steam cylinders, each 3in. diameter and 16in. stroke. This little engine drew a light carriage at ten miles an hour on gravel roads, but it was unfortunately destroyed by fire at the New York Crystal Palace in 1858.

RICKETT. — As we have already said, little or no progress was made in England in steam passenger locomotion between 1840 and 1857. Inventors were actively engaged in railway locomotive improvements, the atmospheric railway gaining a good share of attention. Steamboat propulsion was, during this period, developed very rapidly.

The first to revive the subject of steam passenger carriages was Mr. Thomas Rickett, of Castle Foundry, Buckingham, who completed a road locomotive, which was tested in March, 1858. This engine was capable of traversing any road, and could be steered with precision.

An engine made for the Marquis of Stafford commenced to run during the latter part of 1858, and was fairly successful. Fig. 54 shows this engine, which was carried on three wheels, the two driving wheels behind and one steering wheel in front. The main framing of the engine was formed by a pair of longitudinal iron tanks. The boiler was fixed at the back; the steam cylinders were placed horizontally, one on each side of it, a seat for three passengers being provided in front between the forward end of the boiler and the steering wheel. The crank shaft, as will be seen from the illustration, was placed beneath the seat, the piston rods being coupled on to it in the usual manner. On one side of the crank shaft a small chain wheel was keyed, while a similar wheel of larger diameter was keyed on the driving axle, motion being communicated from the former to the latter by means of an endless pitch chain, as shown. The relative

RICKETT'S STEAM CARRIAGE, 1858.

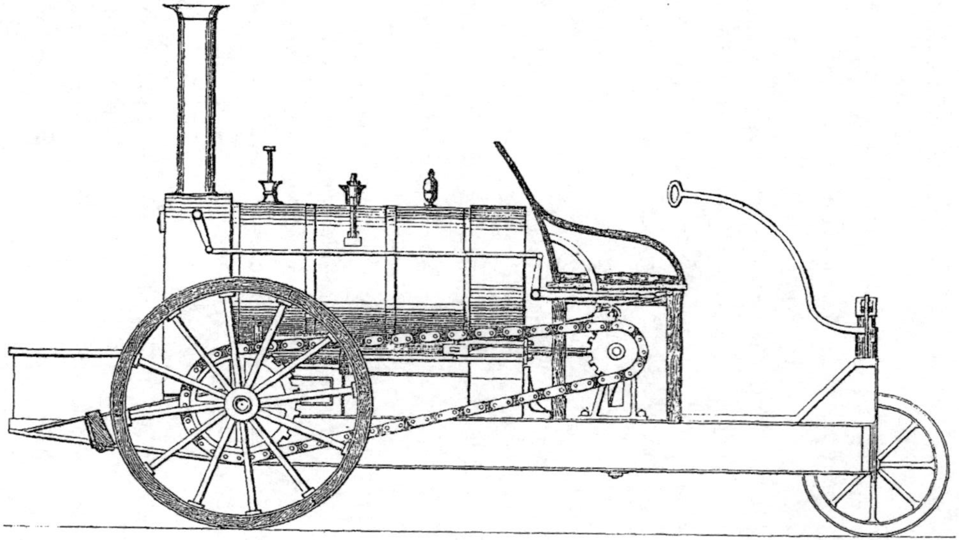

Fig. 54.

sizes of these two wheels were as 1 to 2½. The driving axle was placed as nearly under the boiler as possible, and worked in axle boxes fitted with springs. Behind the boiler was a foot-plate, coal bunker, and seat for stoker. One driving wheel was secured to the axle, the other running loose except when thrown into gear by a clutch. The carriage was steered by means of a lever connected with the fork of the front wheel, which latter passed through a guide in order to allow for the action of the spring. The driver, besides having the steering under his control, was provided with the reversing lever and brake handle, which gave him all necessary command over the carriage. The cylinders were 3in. diameter and 9in. stroke; the working steam pressure was 100lb. per square inch. The driving wheels were 3ft. diameter. The boiler was of the internal flue and return-tube type, and made of steel. The weight of the carriage when fully loaded was only 30 cwt. On good level roads it ran about twelve miles an hour. The *Engineer* for 7th March, 1859, says: "Lord Stafford and party made another trip with the steam carriage from Buckingham to Wolverton. His lordship drove and steered, and although the roads were very heavy, they were not more than an hour in running the nine miles to Old Wolverton. His lordship has repeatedly said that it is guided with the greatest ease and precision. It was designed by Mr. Rickett to run ten miles an hour. One mile in five minutes has been attained, at which it was perfectly steady, the centre of gravity being not more than 2ft. from the ground. A few days afterwards this little engine started from Messrs. Hayes's Works, Stoney Stratford, with a party consisting of the Marquis of Stafford, Lord Alfred Paget, and two Hungarian noblemen. They proceeded through the town of Stoney Stratford at a rapid pace, and after a short trip returned to the Wolverton railway station. The trip was in all respects successful, and shows, beyond a doubt, that steam locomotion for common roads is practicable."

Mr. Rickett built two more engines, substituting spur gearing in place of the pitch chain. One of these carriages was sold to the Earl of Caithness. The cylinders were placed near the passengers' seat, the crank shaft being at the chimney end, near to the main axle, to suit the gearing. The bearings of the driving axle carried the springs, and worked in guides set at an angle from the perpendicular, but at right angles to a line drawn connecting the centres of the two axles, so that the motion of the springs did not materially affect the gearing. There were two sets of spur wheels and pinions, giving proportionate speeds of ten and four miles an hour, so that in ascending hills or traversing rough roads, by throwing in the slow gear, the actual tractive force was multiplied two and a half times. This carriage was intended to carry three passengers, who sat in the front, the stoker being behind. The weight of the carriage, fully loaded, was 50 cwt.*

This locomotive was found to travel exceedingly well, and on good roads attained high rates of speed. The Earl of Caithness, in the carriage just described, travelled from Inverness to his seat, Borrogill Castle, within a few miles of John o' Groat's House. He writes as follows:—" I may state that such a feat as going over the Ord of Caithness has never before been accomplished by steam, as I believe we rose one thousand feet in about five miles. The Ord is one of the largest and steepest hills in Scotland. The turns in the road are very sharp. All this I got over without trouble. There is, I am confident, no difficulty in

* Yarrow, on Steam Carriages, 1862.

THE PERIOD OF SUCCESSFUL APPLICATION. 153

driving a steam carriage on a common road. It is cheap, and on a level I got as much as nineteen miles an hour." The Earl of Caithness brought the trial to a successful result, and ere long steam travelling upon the high roads will be availed of to a large extent. Thus wrote the *Engineer*, but steam passenger travelling on roads has developed very slowly.

"In 1861," we read that "Lord Stafford, now the Duke of Sutherland, had a vertical boiler applied to his carriage," which could not have been any improvement upon Rickett's return-tube horizontal boiler. In 1864, Mr. Rickett supplied an engine for working a passenger and light goods service in

Fig. 55.

Spain, intended to carry thirty passengers up an incline of 1 in 12, at ten miles an hour. The steam cylinders were 8in. diameter, bolted to side frames; the driving wheels were 4ft. diameter. The front of the engine was carried upon a pair of leading wheels placed 2ft. 6in. apart. The boiler could be worked up to 200lb. pressure if required.

We have seen that Mr. Rickett used chain gearing on his earliest engine. His later engines were provided with spur wheels; but he abandoned any form of gearing and made his last engines direct-acting. In November, 1864, he says: "The direct-acting engines mount inclines of 1 in 10 easily; whether at eight, four, two, or one mile an hour, on inclines with five tons behind them, they stick to their work better than geared engines."

Fig. 55 shows a road locomotive and passenger coach constructed by Mr. Rickett, in 1865, similar to the set sent to Spain. The engine would draw a load of 4 tons, at ten to fifteen miles an hour. The steam cylinders were 8in. diameter and 22in. stroke. The driving wheels were 4ft. diameter; the weight of the engine was 6 tons.

These later engines, like the illustration, were simple in construction, without any cog gearing; they were made almost entirely of wrought iron and steel, and were thoroughly well built.

SEAWARD.—In 1859, Messrs. Seaward and Company built an engine to run between London and Leeds, which worked well and ran at a good speed, but from the "oppressive burden of the absurd and disgraceful tolls, it was impossible to carry out the intention with any chance of remuneration to those engaged in the enterprise."[*] The matter remained in abeyance for some time, but was eventually abandoned.

ADAMSON.— Messrs. Daniel Adamson and Co., of Dukinfield, near Manchester, early in 1858 constructed a common road locomotive for Mr. Schmidt, which worked very satisfactorily. The boiler was of the ordinary locomotive multitubular type, 2½ft. in diameter, and 5½ft. long, and intended for a working pressure of 150 lb. per square inch. The engine weighed 56 cwt., and was supported on three wheels. The tank underneath contained 70 gallons of feed water, and the engine was designed for

[*] Young. "Steam on Common Roads," 1861.

running at eight miles an hour. A steam cylinder, of 6in. diameter, was attached to each side of the locomotive. These cylinders actuated a pair of driving wheels, 3ft. 6in. in diameter. Underneath the front of the carriage a single wheel was employed for steering.

Many trials were conducted by Mr. Schmidt in various parts of the country. The following race is recorded in the technical papers for August 30th, 1867. On Monday morning in accordance with previous arrangement, two road steam carriages—one made by Mr. I. W. Boulton, of Ashton-under-Lyne; the other made by Messrs. Daniel Adamson and Co., of Dukinfield—started from Ashton-under-Lyne, at 4.30 a.m., for the show ground at Old Trafford, a distance of over eight miles. The larger engine, made by Messrs. Adamson, was a very well-constructed engine, but the smaller one passed it during the first mile, and kept a good lead of it all the way, arriving at Old Trafford under the hour, having to run slowly through Manchester. The running of both engines was considered very good. On arrival at Old Trafford they tested their turning qualities.

Mr. Schmidt sent this road locomotive to the Havre Exhibition, in 1868, and in accordance with his request a trial of its powers was made by French engineers, and M. Nicole, director of the exhibition. Mr. Schmidt conducted the engine himself, and to it was attached an omnibus containing the commissioners. The engine and carriage traversed several streets of Havre, and mounted a sharp incline. A very satisfactory report was drawn up by M. Ed. Croppi, and sent to several of the journals. Other trips were made to several villages in the neighbourhood of the exhibition, all of which were of a very creditable character.

LOUGH AND MESSENGER. — In 1858, Messrs. Lough and Messenger, of Swindon, designed and erected the little steam road locomotive, illustrated by Fig. 56. This engine, after two years constant running at 15 miles an hour on level roads, and 6 miles an hour up inclines of 1 in 20, was pronounced a perfect success. In the makers' opinion, direct-acting engines were preferable for simplicity's sake; hence the engine illustrated had two cylinders, each 3½in. diameter and 5in. stroke, working direct on to the crank axle. The driving wheels were 3½ft. diameter,

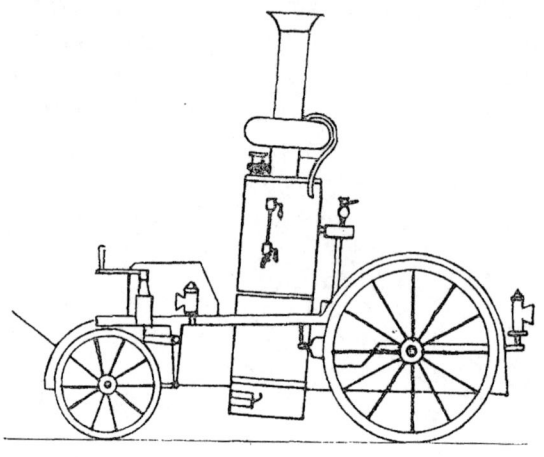

Fig. 56.

and the leading wheels 2ft. diameter. The vertical boiler was fixed on the framing, slightly raking. It was worked at 120lb. pressure. The tanks held 40 gallons of feed water. The boiler was fed by the following means: A small cistern is shown, fixed above the boiler, having communication by a simple pipe from the tanks below. A delivery pipe entered the boiler near the bottom, and a steam pipe from the boiler was carried into the top of the cistern and each pipe was fitted with a cock. The air was expelled from the cistern, and filled with steam, which was accomplished by opening the cocks in the steam and supply pipes, and blowing through into the tank. The steam cock was next shut, as a consequence of which a vacuum was formed in the cistern, and water rushed up from the tank and filled the cistern. The supply cock was next shut, and upon the steam cock being opened, the water fell by its own gravity into the boiler. The total weight of the locomotive was only 8 cwt. Extreme lightness, compared with the power given out, was its chief characteristic. An experienced engineer considered this to be one of the best road locomotives ever turned out.

BACH.—Mr. Bach, engineer, Birmingham, built a small passenger locomotive in 1859, but we are unable to give any particulars of it.

STIRLING. — Mr. Stirling, of Kilmarnock, in 1859, designed a road steamer, in which several neat devices were embodied. All the five travelling wheels were mounted upon springs. A single wheel was used as a driver, and an adjustable arrangement used for causing more or less weight to rest upon this wheel. The leading and trailing wheels were arranged to swivel in concert, in opposite directions, by means of right and left hand worms and worm wheels, whereby a double effect was obtained, and the carriage was made to move in a curve of much less radius than is obtainable with the ordinary steering arrangements.

CARRETT.—Mr. W. O. Carrett, of the firm of Carrett, Marshall, and Co., Sun Foundry, Leeds, made a noted steam pleasure carriage for Mr. George Salt, of Saltaire, which was exhibited at the Royal Show held in Leeds, 1861, and likewise at the London Exhibition, 1862 (see Fig. 57). It had two steam cylinders, 6in. diameter and 8in. stroke. The boiler was of the locomotive multitubular type, made of Low Moor iron. The joints were flanged and welded. It had a copper firebox, and there were 58 copper tubes, 2in. diameter. The boiler was 2ft. 6in. diameter, and 5ft. 3in. long, giving 100ft. of heating surface, and intended for a working pressure of 150lb. per square inch, the test pressure being 300lb. Clark's steam jet for preventing smoke was adopted. The locomotive was mounted upon three wheels, provided with springs. The two driving wheels behind were 4ft. diameter, made of steel, with cast-iron bosses. The leading wheel was 3ft. diameter. We shall refer to the steering gear and special method of attachment to the front wheel presently. The boiler was fed by an injector, as shown, as well as an engine pump. There were seats on the carriage for nine persons, including the steerer and the stoker. The travelling speed was 15 miles an hour; and the weight of the carriage, fully loaded, was 5 tons, about four-fifths of which was carried by the two driving wheels, and one-fifth on the steering wheel. Motion was communicated from the crank shaft to the driving axle through spur gearing in the proportion of five to one. The driving wheels were loose on the main axle, and each was secured upon a hollow shaft revolving upon the main central one; the latter was driven by gearing from the first motion, and carried a pair of central revolving bevel couplings on the inner end of the pipe shaft. The bevel couplings, being free to revolve, became bevel wheels only when there was any differential velocity required to the driving wheels; and thus with both wheels free and always driven, the carriage could turn about in all directions.

CARRETT'S STEAM CARRIAGE, 1862.

Fig. 57.

Mr. W. O. Carrett, after giving the clear description of the compensating gear used on his little locomotive, acknowledged Mr. R. Roberts, of Manchester, as the inventor. There was one point respecting the steering gear worthy of note. The leading wheel, besides being held in its place, says Mr. Yarrow, " by a vertical forked spindle, which passed through guides, had a parallel motion which transmitted any strains taken by the wheel and given out at the axle to a fixed pin or fulcrum on the frame." Thus the brunt of obstacles on the road was taken quite independently, and the steering was left free from any injurious strain. Fig.

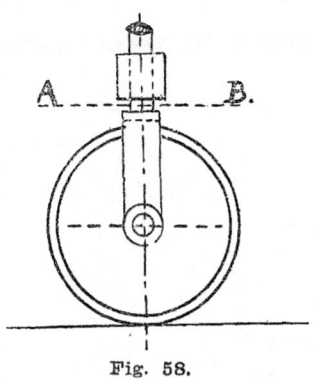

Fig. 58.

58 shows the usual method of connecting the front wheel. The wheel is held in position by a forked spindle, which passes through the guide at the top, which was firmly fixed to the frame; above the guide were the spring and guiding lever. Any obstruction to the forward motion of the carriage had a tendency to break the fork spindle at A B.

Figs. 59 and 60 show the improved method referred to. The wheel in this case is held in position between a forked spindle, but the axle on which the steering wheel revolves is extended on each side, to the ends of which are coupled two links, which connect the front axle and a lever together.

This lever moves on a pin fixed firmly to the framing of the carriage. It will be seen that the wheel can be turned to the right or to the left on the fork spindle as a centre, quite unaffected by the links, which maintain the distance between the axle and the lever under all circumstances the same, and thus the front upright fork spindle is relieved from the injurious strain which we have seen in the former case had a

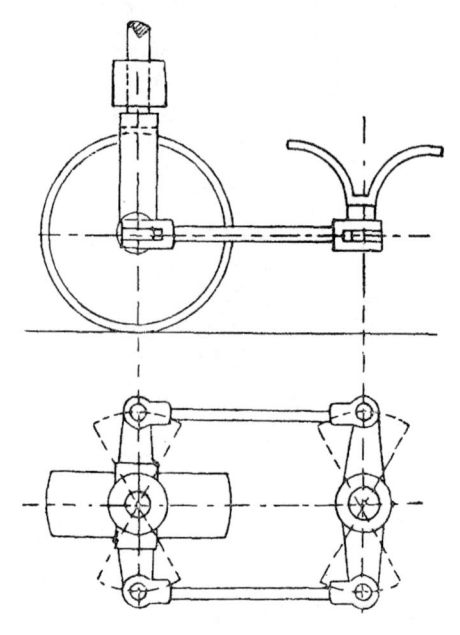

Figs. 59 & 60.

tendency to break the fork at A B.* The following appeared in *Engineering*, 8th June, 1866: "This steam carriage, made by Carrett, Marshall, and Co. (see Fig. 57), was probably the most remarkable locomotive ever made. True, it did little good for itself as a steam carriage, and its owner at last made a present of it — much as an Eastern prince

* Yarrow on Steam Carriages.

THE PERIOD OF SUCCESSFUL APPLICATION. 161

might send a friend a white elephant—to that enthusiastic amateur, Mr. Frederick Hodges, who christened it the Fly-by-Night, and who did fly, and no mistake, through the Kentish villages when most honest people were in their beds. Its enterprising owner was repeatedly pulled up and fined, and to this day his exploits are remembered against him." Hodges ran the engine 800 miles; he had six summonses in six weeks, and one was for running the engine thirty miles an hour. It was afterwards converted into a self-moving fire engine. But the Fly-by-Night was a good job, and deserved a worthier career.

SMITH.—Messrs. Smith Bros., of Thrapstone, had a small self-moving engine at the Smithfield Show in 1861, driven by chain gearing.

YARROW AND HILDITCH. — Messrs. Yarrow and Hilditch, of Barnsbury, near London, showed a steam carriage at the London Exhibition, of 1862, made by Mr. T. W. Cowan, of Greenwich, which is illustrated by Fig. 61. It was designed to carry thirteen passengers, including a steersman and a stoker. A vertical multitubular boiler was employed, 2ft. diameter and 3ft. 9in. high, made of steel. It was fitted with a perforated steam pipe round the top to reduce the chance of priming. The main frame of the carriage was made of ash, 4½in. deep, lined with wrought-iron plates on each side, ¼in. thick. To the outside of the bottom sill were fitted two iron foundation plates, to which the cylinders and other parts were bolted. The cylinders were slightly inclined, as shown, coupled to cranks outside the driving wheels; they were 5in. diameter and 9in. stroke. The driving wheels were 3ft. diameter, both keyed to the main axle and placed inside the framing;

M

YARROW'S STEAM CARRIAGE, 1862.

Fig. 61.

on the outside of the wheels came the axle bearings carrying the springs, and beyond them were keyed the over-neck cranks. The advantages claimed for this arrangement of parts were, "that the gauge of the hind wheels, being considerably reduced, did not necessitate throwing the inner wheel out of gear when turning curves; and, secondly, that direct action was obtained without having the axle cranked, which past experience has shown to be highly objectionable, as the continuous jolting very soon broke them."

In order to keep the axle in its right position, a radius link was employed in preference to axle box guides (we have mentioned elsewhere that the radius link was one of Fisher's inventions), as, by its use, far less friction was caused, and the distance between the shaft and the cylinder was unaltered, and the action of the slide valves was unaffected by the vertical play of the springs. One end of the link was connected to the axle bearings, and the other to a provision on the foundation plate. The working parts were covered to protect them from dust. All the parts of the engine were made as light as possible, steel and malleable cast iron being freely used. When fully loaded the carriage weighed 2½ tons, the greater portion of which was borne by the driving axle. The carriage performed well on the road.

LEE.—Mr. Lee, of Leicester, sent a three-wheel road locomotive to the London Exhibition of 1862. The driving wheels were 6ft. diameter, the cylinders 4½ in. diameter and 9in. stroke. Motion was transmitted from the crank shaft to the main axle by spur gearing, in the proportion of eight to one.

HAYBALL.—A quick-speed road locomotive was made by Mr. Charles T. Hayball, of Lymington, Hants, in 1864. The whole of the machinery was mounted upon wrought-iron framing, 4in. deep, and ¾in. thick, and supported by three wheels. The two driving wheels had an inner and an outer tyre, and the space between the two hoops was filled up with wood to reduce noise and lessen the concussion. The two steam cylinders were bolted to the top of the frame, each 4½in. diameter and 6in. stroke. A vertical boiler, 2ft. 2in. diameter, and 4ft. high, working at 150lb. pressure, was used. The ratio of the gearing was 10 to 33. The crank shaft was placed above the main axle, both being fixed in a sliding bracket, keeping the shafts the right distance apart to suit the gearing, both sliding up and down owing to the action of the springs. The carriage ran up an incline of one in 12 at 16 miles an hour, and travelled 4 miles in 14 minutes up hill and down with 10 passengers on board. Hayball says "he has run 700 miles without a derangement, and travelled up an incline a quarter of a mile long, 1 in 10 and 1 in 12, and when nearly at the top he turned at right angles." When loaded with 12 passengers the carriage weighed under two tons. A bar was fixed in front of the vehicle for removing large loose stones on the road out of the way, a device of questionable utility.

WILKINSON.—In 1862, Mr. Wilkinson, of Ashford, Kent, possessed a small steam carriage with two direct acting cylinders each 4½in. diameter and 12in. stroke. The driving wheels were 5ft. diameter. Steam of 120lb. pressure was supplied by an ordinary vertical boiler.

TANGYE.— Messrs. Tangye, of Birmingham, in 1862, constructed the road locomotive as per Fig. 62.

THE PERIOD OF SUCCESSFUL APPLICATION.   165

The illustration (kindly lent by Messrs. Tangye) shews the engine so clearly that little descriptive matter is needed. The centre of the carriage afforded ample space for seating six or eight persons, whilst three or four more could be accommodated in front, one or two of whom performed the duties of driver and steersman. The stoker was of course stationed at the boiler behind. The driver who sat in front had full control of the stop valve and reversing lever, so that the engine could be stopped, or reversed by him as occasion required, whilst by means of a very powerful and well arranged foot brake at his command, he was able to bring the carriage to a standstill in an incredibly short time and distance. The management of the whole engine was so simple that the most unskilled persons might have undertaken it without the slightest fear of accident. The speed of 20 miles an hour could be readily attained, and the engine with its load ascended the steepest gradients with perfect ease and safety.

The body of the carriage was made of iron and supported on steel springs of great flexibility, the motion over the roughest roads being smooth and easy. The length of the carriage was 16 ft., and the width 5 ft. 9 in. The two cylinders were $5\frac{1}{2}$ in. diameter and 11 in. stroke. The cylinders were neatly lagged, and with the guides were protected by an iron casing as shown in the illustration Fig. 62.

The driving and steering wheels were each 39 in. diameter, and 2 in. on the face, made of wood and strongly tyred with iron. The whole vehicle was remarkably compact and simple in construction, and the working parts were few in number and not liable to derangement. The vertical boiler had a copper firebox, and contained 100 brass tubes enabling steam to be raised in a few minutes.

TANGYE'S ROAD LOCOMOTIVE, 1862.

Fig. 62.

No inconvenience was felt from the heat of the boiler by persons seated in the body of the carriage, it being partly surrounded by the feed water tank, the steam was dried before entering the cylinders by passing through a conical coil of pipes in the smoke-box. When burning coal, a small jet of steam was employed to introduce air above the fire, and was found to be very effectual in preventing smoke. The locomotive carried sufficient fuel and water for a journey of 20 miles.

From Mr. Richard Tangye's interesting autobiography recently published, we quote the following respecting this steam carriage :—" About 1862 the subject of providing 'feeders' in country places for the main lines of railway came again into prominence. Branch lines had been proved to be unremunerative from their great cost in construction ; and amongst other systems proposed was that of light, quick-speed locomotives for carrying passengers, and traction engines for the conveyance of heavy produce and other goods. We determined to construct a locomotive of the former class, and succeeded in making a very successful example, with which we travelled many hundreds of miles. The carriage occupied no more space than an ordinary phæton ; when travelling at over 20 miles an hour, the engine was easily managed and under perfect control. Fig. 63 shows the 'Cornubia' on the village green.*

" Great interest was manifested in our experiment, and it soon became evident that there was an opening for a considerable business in these engines, and we made our preparations accordingly, but the 'wisdom' of Parliament made it impossible. The squires became alarmed lest their horses should take fright ; and although a judge ruled that a

---

* This block (kindly lent by Mr. Richard Tangye) is one of the illustrations specially prepared for the autobiography *One and All*.

horse that would not stand the sight or sound of a locomotive, in these days of steam, constituted a public danger, and that its owner should be punished and not the owner of the locomotive, an Act was passed providing that no engine should travel more than four miles an hour on the public roads. Thus was the trade in quick speed locomotives strangled in its cradle, and the inhabitants of country districts left unprovided with improved facilities for travelling." In no single instance did Messrs. Tangye's carriage cause an accident attributable to horses. At one time a countryman

Fig. 63.

tumbled out of his cart from fear that his horse might bolt, but the latter was wiser than his master, for he stood quite still.

BOULTON.—Mr. Thomas Boulton, in August, 1867, says: "I ran a small road locomotive constructed by Mr. Isaac W. Boulton, of Ashton-under-Lyne, from here through

Manchester, Eccles, Warrington, Preston Brook, to Chester, paraded the principal streets of Chester, and returned home, the distance being over ninety miles in one day without a stoppage except for water. I believe this to be the longest continuous run on record ever accomplished by any road locomotive within twenty-four hours." But in this Mr. Boulton was mistaken. We have stated in a previous article that Hill ran from London to Hastings and back in one day, a distance of one hundred and twenty-eight miles. Boulton's engine had one cylinder $4\frac{1}{3}$in. diameter, and 9in. stroke. The boiler was worked at one hundred and thirty pounds pressure per square inch. The driving wheels were 5ft. in diameter. The driving and the single steering wheels were provided with springs.

Two speeds were obtained by means of two trains of spur gearing between the crank shaft and the counter shaft, the motion of the counter shaft was transmitted to the axle by a pitch chain; the ratios of the gearing were $6\frac{1}{2}$ to 1, and 11 to 1. During the trip recorded above, six persons were carried all the distance, and sometimes there were eight and ten passengers. This is the little engine which ran a race with Mr. Adamson's locomotive, recorded in a previous chapter.

GOODMAN.—A small road locomotive was made about this time by Mr. Goodman, of Marshall street, Southwark, London. It was worked by a pair of direct-acting engines, coupled to the crank shaft in the usual manner. A chain pinion on the crank shaft transmitted motion to the main axle through an endless pitch chain, working over a chain wheel of larger diameter on the driving shaft. The steering gear was arranged like Carrett, Marshall, & Co.'s engine, so that the jerks or strain on the steering fork were reduced to a minimum. The smoke from the boiler was conducted by a flue placed beneath the carriage, and issued out behind. This neat little conveyance ran ten to twelve miles an hour.

ARMSTRONG.—Sometime in 1868, Mr. Armstrong, of Rawilpindee, Punjab, India, made a neat little steam carriage, having two steam cylinders, each 3in. diameter, and 6in. stroke. A trunk was used, so as to do away with slide bars. A separate stop valve was fitted to each cylinder. The boiler was 15in. diameter, 3ft. high, and was worked at 100lb. steam pressure per square inch. The carriage travelled at twelve miles an hour on the level road, and up an incline of one in twenty at the rate of six miles an hour. The driving wheels were 3ft. diameter. The engine had been running more than a year when the above particulars were forwarded by Mr. Armstrong to *The Engineer*.

MODERN PERIOD.

THOMPSON.—We have now arrived at a period when steam road locomotion was rapidly developed. Engines of a special type for either fast or slow speeds were made in great numbers, and sent to all quarters of the globe. Much of this progress was due to the inventive ability and indomitable energy of Mr. R. W. Thompson, C.E., of Edinburgh, who not only had a considerable number of these road steamers made at Leith under his personal superintendence, but, because of the briskness of the trade which quickly sprang up, three or four firms took up the manufacture.

Mr. R. W. Thompson was born at Stonehaven in 1822. Early in life he was sent to the United States of America to be made a merchant of, but he disliked the calling, and returned to England when sixteen years of age. He spent two years of his life making experiments in chemistry and electricity, interspersed with engineering schemes. He was next apprenticed to engineering, at Aberdeen and Dundee, filling up his spare moments during this period in inventing a rotary engine, &c.

After serving his apprenticeship during which he made rapid progress, he was employed by the Stephensons. In 1844 he commenced business on his own account. Two years later he conceived the idea of applying indiarubber tyres to ordinary conveyances. We read in vol. xlv. of the *Mechanics' Magazine*, that noiseless tyres had been applied to a brougham which was running in the London parks, the invention of Mr. Thompson. Indiarubber at that time was scarce and badly made, so the invention brought in poor returns. Moreover, the powerful railway companies in due course adopted the tyres to the platform handcarts, and paid him no royalty. Thompson sent in a plan for the 1851 Exhibition, which received some attention; and a fountain pen of his invention was sold inside the exhibition.

In 1862 he had settled in Edinburgh. The portable crane was one of his most useful inventions. A traction engine being required for use in Java, from whence Thompson had recently returned, he commenced to design one in harmony with his own ideas, which resulted in his invention of the indiarubber tyres for the wheels, and the "pot" boiler, in 1867, which made his name famous. Although numerous inventors had cherished the idea of applying indiarubber or other soft substances, covered with leather, &c., to the tyres of road locomotives, before 1867, they having no doubt received the inspiration from his noiseless tyres in 1846, yet he was the first to put the idea into practical shape.

In December, 1867, a small road locomotive having a "pot" boiler and vulcanised rubber tyres to the wheels was being tested, and the newspapers pronounced the engine to be "in advance of everything which had preceded it." The steam cylinder was 5 in. diameter, and 8 in. stroke. The engine was mounted upon three wheels, all of which were fitted with rubber tyres, the driving wheel tyres being 12 in. wide and 5 in. thick. Numerous trials were run with this engine, drawing a large omnibus behind, at the rate of 10 to 12 miles an hour. It was said: "Mr. Thompson intends to run the engine over to Glasgow by the road, for shipment to Java, where it is to be used for travelling between two towns, about 40 miles apart, taking in tow a large omnibus full of passengers, or trains of wagons, at the speed which has already been acomplished in the trials which have been made in the neighbourhood of Edinburgh." We illustrate by Fig. 64 one of Thompson's road steamers, made in 1868 by Messrs. Tennant and Co., of Leith, for use in the Island of Ceylon.

From the illustration it will be seen, the horizontal engine and vertical pot boiler were mounted upon a wrought-iron

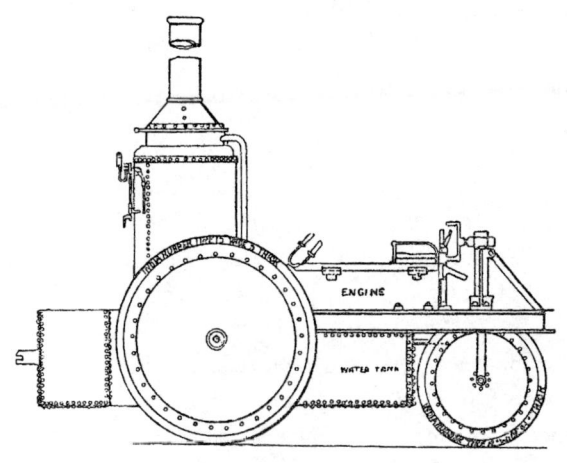

Fig. 64.

frame of channel iron, presenting a neat and compact appearance. This engine was subjected to some severe trials previous to its shipment for Ceylon. We are unable to afford the space to give particulars of a tithe of the trips made by Mr. Thompson with his road steamers. However, in 1869 some trials were made with two 6-horse engines, under Mr. Thompson's directions, which exhibited their tractive power and speed in a remarkable manner. "One of the 6 ton road steamers was harnessed to four wagons of pig iron—weight of iron and wagons, 34 tons—which it drew without an effort or any stoppage from the foot to the top of Granton Road, a distance of a quarter of a mile, with inclines of one in eighteen. Arrived at the top, it turned with its train and ran back to its starting point. It may be pointed out that the drawing of 34 tons, besides the engine's own weight, up one in twenty, is equal to drawing 100 tons or more on a level road. The other road steamer was attached to an omnibus which conveyed a party of gentlemen from Granton to Leith. The distance is two and three-quarter miles, and the journey was performed at the rate of over eight miles an hour, that being the highest speed at which it was deemed safe to run through a town." This road steamer had been built specially for omnibus traffic, and was exceedingly light and compact. One morning a road steamer was taken down on to the sea sands at Portobello, and ran up and down there at the rate of ten miles an hour, the rain pouring all the time in torrents. A road steamer was employed at Aberdeen to draw a 15 ton boiler on a 5 ton wagon through some of the streets at three miles an hour. An engineer writes: "It is certainly a feat for a 5 ton engine to drag out a 20 ton load and climb gradients of one in twenty with single gear. We had all Aberdeen turned out as we passed. It was an unusual sight to see the infant 'Hercules' in front of the great boiler, 30 feet long and 7 feet in diameter, bowling along with it like a plaything at its tail, whilst the plaything itself shook the houses again as it danced over the rough causeway."

It would be amusing to quote the foolish statements made respecting the unprotected vulcanised rubber tyres; indeed a great deal of harm was done to the cause of indiarubber tyres by the inconsistent statements of some of their promoters, who invested them with almost marvellous capabilities. One advocate remarked that because the soft tyres resembled the feet of the elephant and the camel, which have large soft cushions in hard hoofs, and as no other animal can bear so much walking over hard roads as they can accomplish, therefore these tyres would pass over newly-broken road metal, broken flints, and all kinds of sharp things without leaving a mark on the rubber. And we were

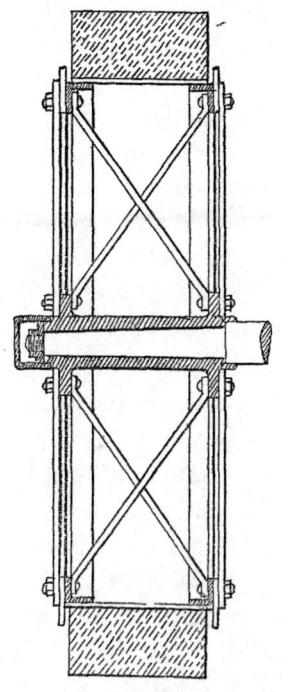

Fig. 65.

repeatedly nformed that the tyres were not affected by heat, cold, or moisture, and were durable beyond all conception; and yet, in the face of all this claimed durability, no end of schemes were being proposed and patented for protecting the surface of the tyres from injury. As one writer nicely puts it, when noticing one of the methods of attaching metal shoes round the tyres : " Considering how much has been said concerning the everlasting properties of the indiarubber tyres, it is curious that so much ingenuity should be expended in affording them protection."

In January, 1870, Mr. Thompson sent out an eight horse road steamer to India, which, though not a success (says Mr. Crompton), proved that the rubber tyres were not affected by climate, and that the engine was handy and manageable. Four larger engines were eventually ordered by the Indian Government of which particulars follow. At the Royal Show at Oxford, July, 1870, two road steamers with india-rubber tyres were shown, running about the ground, " twisting, turning—we are inclined to say waltzing—and disporting themselves generally in a manner suggestive of what a pair of gigantic steam kittens or other frolicsome animals might do. One engine was tried without the steel chain armour around the wheel tyre, and on the strain being thrown on one wheel the tyre on that wheel snapped right across."*

Fig. 65 shows a section of Thompson's original rubber tyred wheel, which was constructed of wrought iron plates, strengthened by angle iron and diagonal stays, with low flanges on each side to keep the tyres in place. The periphery of the wheel was polished, and then drilled all over with ⅜in. holes. The elastic tyre was made a little less in diameter than the wheel, and being stretched in order to get it on the wheel, had a tendency to contract, which helped to keep it in its place. The boss or nave of the wheel was made of cast iron. The wheel ran with the indiarubber in contact with the ground. To prevent the wheels from slipping on soft and damp roads, the chain armour was introduced. Fig. 66 shows a part side elevation of a rubber-tyred wheel with

* *Engineering*, Vol. 10.

MODERN PERIOD. 177

the chain armour applied. Fig. 67 shows a section of the same wheel. This species of armour consisted of steel plates or shoes, joined together by flat links of malleable cast iron, and was a constant source of annoyance in practice, owing to the breakage of the link pins, and the difficulty of keeping

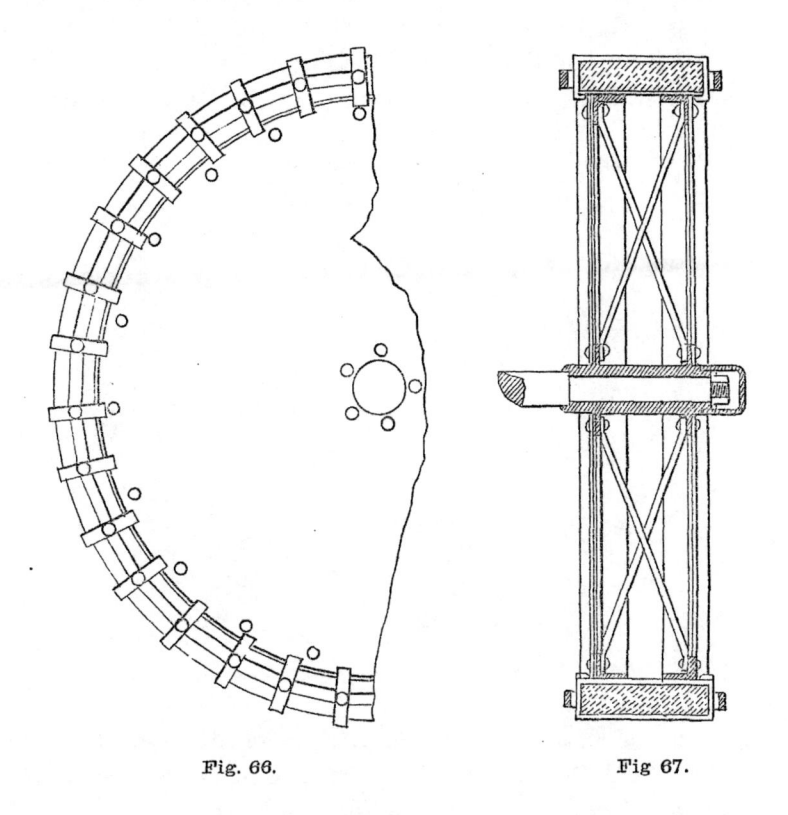

Fig. 66.   Fig 67.

the tyre in its place unless the shoes were very tight. When an indiarubber tyre is working without shoes, at a speed of 8 miles per hour, there is a much greater amount of india-rubber on the leading side than on the following side of the wheel. On the leading side the excess of indiarubber ac-

N

commodates itself by 'bagging out,' as shown by Fig. 68,* while in the rear it is in a state of tension, and tightly grips the iron wheel. From this action the indiarubber tyre is continuously working round with a reverse motion to that of the drum. The rate of this motion depends upon the tightness with which it was originally stretched, its density, its thickness, and the weight of the wheel. If the wheel is lightly loaded, the tyre will scarcely move, while if it is heavily

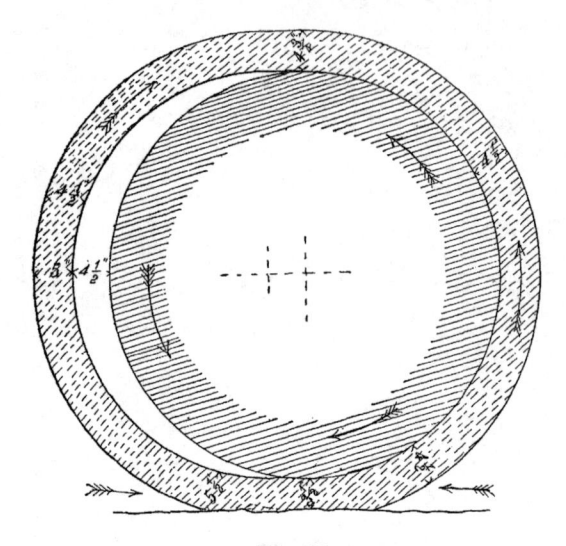

Fig. 68.

compressed, a great portion of it is rolled out towards the front, and the amount of the reverse action becomes very great. Under ordinary circumstances the tyre will move once round the drum in from 30 to 40 revolutions. It is evident that friction must take place in the indiarubber tyre, from its contrary rotation round the iron drum, and also from the continuous change of form it undergoes.

* Mr. Head's Rise and Progress of Steam Locomotion on Common Roads.

It is self evident from the above remarks that indiarubber tyres, to be successful, should be relieved of all driving strain.

In spite of the bad name, and whatever drawbacks the indiarubber tyres may possess, it is impossible to ignore the following facts, which must be placed to their credit. They act as an excellent spring, and are placed where the spring should be situated—that is, in the nearest point to the road, thus saving the engine from a great amount of wear and tear and rough usage. They are perfectly noiseless. Owing to their flexibility they always possess a regular amount of surface of adhesion, which on paved roads is almost indispensable. The injury to the road may be said to be *nil*, for there has never been any complaint from the road authorities of any damage done by them; and they are one of the means devised for enabling a road locomotive to travel over the highway at a sensible speed, say 7 to 10 miles an hour. Another advantage of the rubber tyre is mentioned by Mr. John Head. "On good macadam its resistance is more than that of the rigid wheel, and on a rough or newly-metalled road, owing to its great surface, it does not sink below the tops of the stones, while the rigid wheel consumes a great amount of power from sinking into the surface of the road with a crushing and grinding action." The great cost of the rubber tyres had, no doubt, much to do with their ultimate disuse. The failure of the indiarubber tyre at the trials at the Royal Show, at Wolverhampton, in 1871, owing to the slippery state of the land after the excessive rains, is very well known. "The clay was spurted up from under the wheels, and entered between the indiarubber and the rim of the drums, and so lubricated the parts that there was a revolution of the iron rim within the indiarubber." Thompson's tyres answered well for regular road purposes. The experiments carried out by Mr. Crompton, in India, proved: "(1) That upon the level roads of India, traction engines can be relied on to work a service of trains with great regularity and at a fair speed, and that passengers can be conveyed at eight miles an hour. (2) That the rubber tyres, as used in such running, are of great service in reducing the cost of the ordinary engine repairs, and in giving uniformity of adhesion, without in the least degree damaging the surface of the roads."

Such was the demand for Thompson's road steamers between 1870 and 1873, that Messrs. Tennant and Co., of Leith, could not make them fast enough. Engines of this type were made by Messrs. Robey and Co., of Lincoln; Messrs. Ransomes, Sims, and Head, of Ipswich; Messrs. Charles Burrell and Sons, of Thetford; and others, particulars of which follow. Thompson's pot boiler was not a success; it was abandoned in favour of the Field vertical, or locomotive multitubular type of boiler. The indiarubber tyres, with suitable protection, continued to be used until the time of Mr. Thompson's death, which occurred on the 8th of March, 1873, he being in his 51st year. Although his two chief inventions were not a thorough success, yet he paved the way for other schemers, and by his efforts steam locomotion on common roads was rapidly advanced.

TENNANT.—Messrs. Tennant and Co., of Leith, made a large number of road steamers for Mr. Thompson. Some of these, intended for passenger service, are already referred to; but they also made a considerable number of engines for heavy haulage purposes, one of which is illustrated in Mr. John Head's paper "On the Rise and Progress of Steam Locomotion on Common Roads," 1873.

TODD.—In 1869, Mr. Todd, of Leith, made a neat little road locomotive mounted upon springs, having two steam cylinders, each 2½ in. diameter and 4 in. stroke. The two driving wheels were 4 ft. diameter; the engine would run one hundred miles in ten hours.

Three years later Mr. Todd designed and constructed a special road locomotive to run mails and passengers at high speeds, in such places abroad where there were good main roads, between important stations not connected by railways. The engine was carried on two driving wheels, having springs, 6 ft. centres, and two leading wheels, each mounted upon volute springs. The steel springs were likewise supplemented by rubber washers, so that no trace of vibration remained. The boiler was of the locomotive multitubular type, intended for working at 180 lb. pressure. Two cylinders, 10 in. diameter and 10 in. stroke, were connected to the crankshaft in the usual manner: spur gearing having a ratio of four to one connected the crankshaft and a countershaft together, motion being transmitted from a disc on the countershaft to the main axle by side coupling rods. Fisher's parallel rods were also used for connecting the axle boxes of the countershaft shaft and the main axle together. By this arrangement the axle can bear any amount of vertical play without the working spur gearing being at all affected. All the wheels had solid wood discs and iron tyres. The engine would, on a suitable highway, take a load not exceeding seven tons a distance of one hundred miles in ten hours. Mr. Todd offers the following remarks respecting the rigid wheels: "It has been stated that it is impossible to run a rigid-wheeled engine twenty miles an hour on an ordinary macadamised road. The statement is simply not correct, for this has frequently been done. A rigid wheel carrying four tons, with a 50 in. bearing spring and rubber washers, runs with smoothness on macadam at twenty miles an hour; and again, another rigid wheel, also loaded with four tons, and fitted with two volutes and rubber washers, runs as smoothly at twenty miles an hour on a good road as an ordinary railway carriage. The fact is, rigid wheels have got a bad name as being rough to ride on, simply because they are almost universally used in road locomotives without any springs at all, or at best with springs so stiff as to almost prevent any motion of the axle; and thus the wheels are blamed for what is most obviously the fault of the designer." In 1872, Mr. Todd constructed a steam omnibus for running between Leith and Edinburgh.

NAIRN.—In 1870, Mr. Nairn, of Leith, designed a steam omnibus for service in Edinburgh; it had one leading and two driving wheels, fitted with willow wood on the faces to deaden noise. Three cylinders, each 5¼ in. diameter and 10 in. stroke were coupled direct to the driving axle, which was placed well forward to relieve the weight on the steering wheel; the carriage was mounted upon springs 4 ft. 6in. long, and rubber washers. A Field boiler, 27 in. diameter and 6 ft. high, was used, the funnel from which ran along the top of the 'bus The boiler was sufficiently large, so as to dispense with the exhaust blast in the chimney. The waste steam was allowed to escape into a box, which made it noiseless. The omnibus weighed seven tons, and seats were provided for eighteen inside and thirty-two outside passengers. With fifty persons on board, the engine ascended a hill, one in sixteen, at eight miles an hour, and ran on the level road at sixteen miles an hour. An Edinburgh 'bus proprietor took up this machine, and drove a heavy trade for a time.

During the same year Mr. Nairn built a very neat-looking road steamer of eight horse power, mounted upon three wheels, each trailing wheel being driven through the medium of a friction brake lined with wood. These brakes, it was urged, only required once tightening in a hundred miles run.

These friction brakes were adopted so as to dispense with compensating gearing. The engine could be manœuvred easily. A Field vertical boiler was used on this engine, supplying steam to two cylinders, each 6½ in. diameter and 10 in. stroke ; two speeds of steel gearing. The two driving wheels were 5 ft. 6 in. diameter and 18 in. on face, a section of which is shown by Fig. 69, on this page. It will be seen that the tyre was made of hemp and coir rope. He sometimes used cork and flat pit ropes coiled round a drum or wheel, and protected by shoes on the outer surface. " It was found in practice that when the rope was coiled loosely enough to

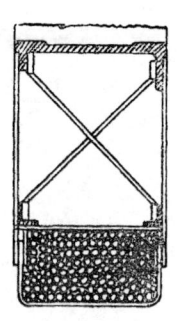

Fig. 69.

flatten at the tread of the wheel to the same extent as india-rubber, the resistance became very great through the tyre being soft and not elastic, and the rope was soon destroyed. If the coils were wound round as tightly as possible, the tyre became almost rigid, and the wheel was not more efficient than an iron one."

In 1871 Mr. Nairn brought out his self-contained steam omnibus, the 'Pioneer.' Three cylinders were coupled to a pair of 40 in. diameter driving wheels. The funnel from the boiler passed backwards under the seats of the outside passengers, and emerged behind over the conductor's head. The main axle was mounted on flat bearing springs, supplemented by indiarubber washers 4 in. thick. During the summer of 1871, the 'Pioneer' ran for hire between Edinburgh and Portobello, a distance of three miles, and from eleven to twelve runs were made per day. Seats were provided for fifty passengers, and when fully loaded the omnibus weighed ten tons, travelled at twelve miles an hour, was under perfect control, and was successful. An engineering writer says: "Having travelled on business from Edinburgh to Portobello, I have had frequent opportunities of observing the working of this steam omnibus, and it is doing exceedingly well. No horse-drawn 'bus is more under control than this one; its safety and capabilities of doing excellent work are beyond cavil, and invite investigation. Its general construction is a great step in advance." Nairn, like the rest of the locomotionists was more than once summoned before the magistrates and fined for breaking the unreasonable road locomotive laws. In 1872, Mr. Nairn was having a road locomotive built to his design by Messrs. J. and T. Dale of Kirkcaldy, for shipment to New Zealand.

KNIGHT.—Mr. J. H. Knight, of Weybourne House, Farnham, had a little steam carriage made in 1868; it was capable of carrying three passengers at a speed of 8 miles an hour on good roads, while it easily mounted the hills in the neighbourhood of Farnham. The carriage was propelled by two cylinders, each 5 in. diameter and 12 in. stroke, the power being transmitted from the crank shaft to the driving axle by chain gearing, the ratio of which was four to one. The vehicle was mounted on four wheels; the two driving wheels were 4 ft. diameter. The leading wheels 2 ft. 8 in. diameter. The carriage was supplied with a brake, a donkey pump, and a feed water heater.

CATLEY.—A very neat steam wagonette was made by Mr. Catley, of York, in 1869. The two cylinders were 2⅜in. diameter, and 5¾in. stroke. Spur gearing with a ratio of three to one was used for transmitting the motion of the crank shaft to the main axle. The driving wheels were 4ft. diameter; one wheel was keyed to the axle, the other wheel was loose on the shaft, which caused the carriage to turn the sharpest corners easily. A vertical boiler, 1ft. 6in. diameter and 3ft. high, was mounted over the driving axle, and a pressure of 120lbs. per square inch was used. Four passengers could be carried at a fair speed. The weight of the wagonette empty was 15cwt.; it was mounted on good springs, fitted with a brake within easy reach of the steersman, and nicely equipped for service. Two water tanks contained a supply of feed for a five mile run, while coal enough could be taken for a 20 mile journey.

ROBEY.—Soon after the introduction of Thompson's rubber tyres, the orders for road steamers, both for heavy haulage purposes in Great Britain, and fast speed travelling in foreign countries, came to hand so rapidly that Messrs. Tennant and Co., were unable to meet the demand. Messrs. Robey and Co., of Lincoln, were one of three well-known firms who took up the manufacture of these engines for Mr. Thompson. In 1870 the Lincoln firm made a large road steamer, called the 'Advance,' for Woolwich Arsenal, fitted with rubber tyres and "pot" boiler.

Fig. 70 shows the engine clearly, while Fig. 71 gives a sectional elevation of the road steamer to a large scale. The engine was of the vertical type, having two cylinders, each 7¾ in. diameter and 10 in. stroke; the crankshaft was 3 in. diameter, the countershaft was driven by spur gearing from the crankshaft as shown.

This gearing was keyed fast to both shafts, so the countershaft was always running when the engine was in motion, the pump being driven from this shaft. For

Fig. 70.

obtaining two travelling speeds, both the crankshaft and the countershaft were fitted with spur pinions sliding on fixed keys, either of which could be made to gear

ROBEY'S ROAD STEAMER, 1870 (Sectional View).

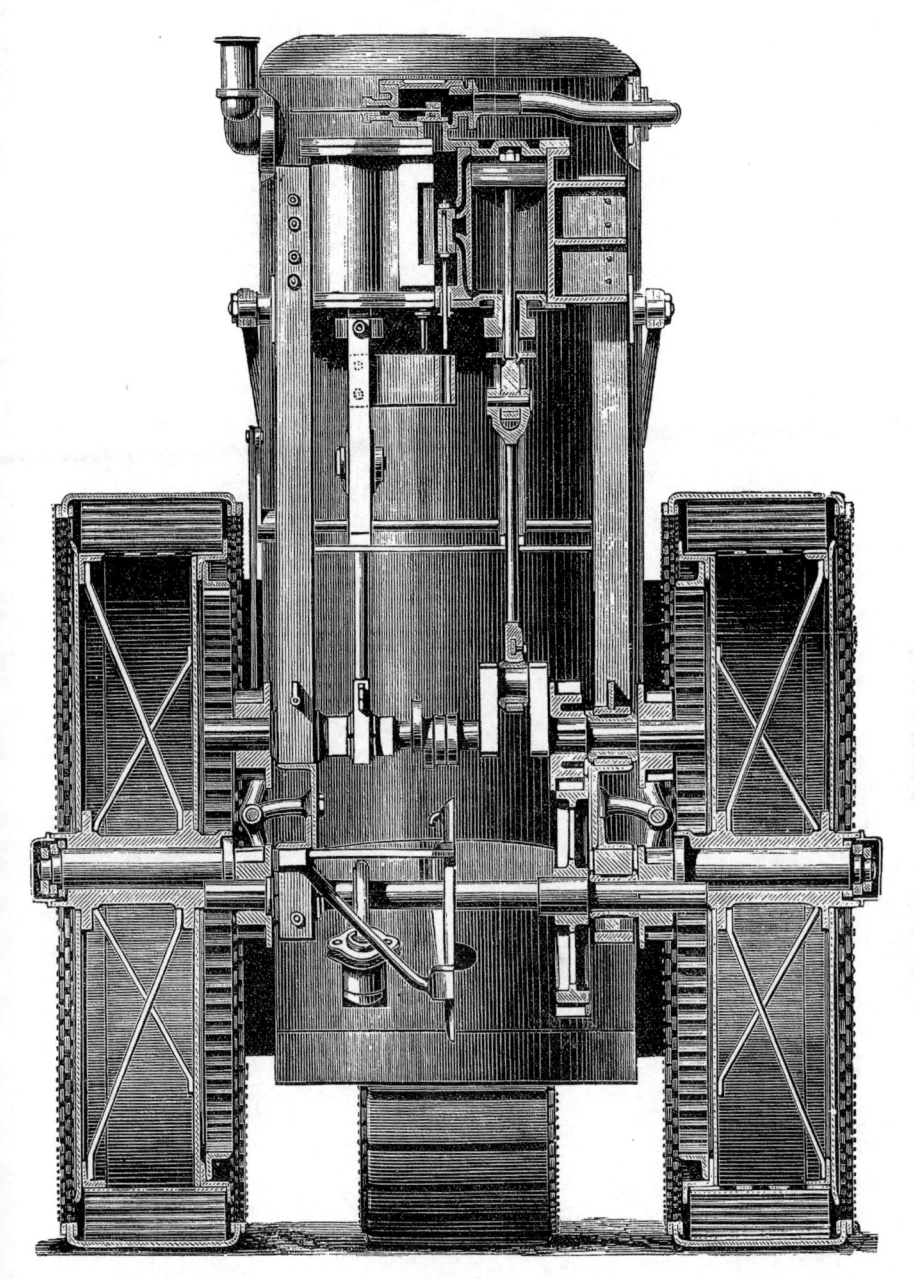

Fig. 71.

with a spur ring on the road wheel. As this arrangement of gearing was adopted on all Thompson's engines, we give diagrams to illustrate it. See Figs. 72 and 73. It will be obvious that if the pinions on the crankshaft

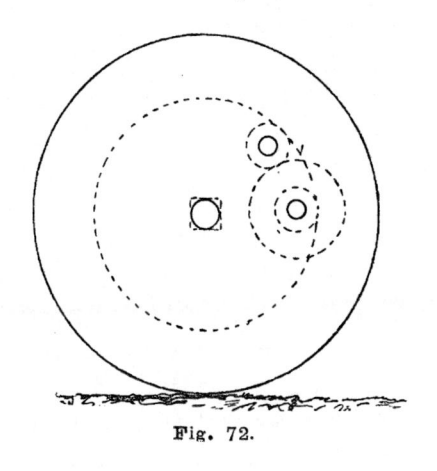

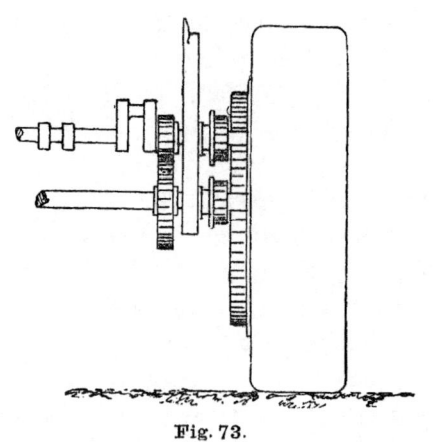

Fig. 72.

Fig. 73.

are in gear with the travelling wheels, the engine will travel at a fast speed, and when the countershaft drives, the slow speed is obtained. For turning sharp corners, one of the pinions could be readily thrown out of gear. The driving wheels were 6 ft. diameter, and the leading wheel 4 ft. diameter; the rubber tyres were 5 in. thick. An interesting test was carried out with this engine in December, 1870, as follows :—" The first experiment was to show the adaptability of the road steamer for passenger traffic, and for this purpose a break and an omnibus were attached at Messrs. Robey's works in Canwick Road, and, with a load of 45 passengers, proceeded at a smart pace—not less than six miles an hour—along the level and slighter inclines making two sharp curves, and running over a very awkward short and steep hummock, formed by the iron bridge over the Witham, in the route to the Lindum Hill, the steepest gradient on which—1 in 9—it did at a speed of between four and five miles. The 'Advance' then turned on the hill-top with its train in a circle, the inner diameter of which was about 18 feet. The run down the hill, which is a full half-mile, was made at times at a great speed, the crowd of sight-seers all running to keep up with it. At other times the engine was checked and brought almost to a stand upon gradients of every degree of severity up to the extreme one.

After the return to the works the carriages were unhooked, and a train of two four-wheeled trucks, weighing three tons each, and carrying loads of two tons of deals, in all ten tons attached. With this the 'Advance' proceeded along the macadamised turnpike road, up the Canwick Hill, the heaviest gradient of which is one in eight. The purpose of this experiment was to show the capacity of the Steamer for drawing heavy loads on ordinary roads, and the test was a severe one, in consequence of the surface coating of slippery mud. The hill nearly three quarters of a mile on the rise, was handsomely got up at the rate of two and a half miles, the engine with its train—in all 46 ft. in length—turning within circles of the following dimensions: Exterior diameter of engine track 24 ft. 6 in., and exterior diameter of truck tracks 30 ft. The journey down hill was again literally at the run, the trucks having each a slipper on one wheel; the speed kept up in the descent was quite six miles, the control and steering being perfectly kept in hand by Mr. Stanger, the instructor of engine-driving at Woolwich and Aldershot, who took charge personally throughout the day, and handled the engine in a masterly manner. The 'Advance' started with this load at 1.45 p.m., and was stopped at the works on the return at 2.15 p m., the journey including two stoppages, one to put on and one to adjust the slipper skids, occupying exactly half an hour. Part of the distance was over newly-metalled road, of which, however, the wheels made not the least disturbance— it being one of the characteristic features of these engines that they do no damage whatever to the roadways, and, indeed, they do not injure grass lands, as was convincingly shown in the runs which were next made in the large meadows expressively known locally as the "Cow Paddle," but otherwise the South Common. This land, lying low, was exceedingly spongy, and the 'Advance' in places sunk at times from three to four inches in some of the soft places. The purpose, indeed, of the third experiment was to show the capacity of the Road Steamer for going over soft land, and the "Cow Paddle" was an undoubted test of this qualification. A measured mile was stumped out, and the engine run round the course, doing the whole distance in seven minutes, notwithstanding one-third was over ground so soft and wet that the engine worked over that portion with an average sinking of at least two inches. Several very short turns were made by the 'Advance' upon this grass land at sharp speed, in one case the inner diameter of the wheel tracks being only seven feet, measured across the second or innermost spiral. In making another circle the 'Advance' passed over a hollow ten inches deep in seven feet of length, one wheel sinking into the soil five inches below this; and yet there was no arrest, not even instantaneous, in the progress of the engine. A speed of nine miles an hour upon such ground as this is a wonderful accomplishment, and the steadiness with which the 'Advance' worked upon ground of the most uneven nature shows clearly the great value of these machines for the roughest locomotive work. The indiarubber tyres are guarded by an outside band of steel plates, 18 in. broad and 5 in. deep, with intervals of 1¼ in. between them. At almost all times there were four of these plates bearing firmly on the ground through the elasticity of the rubber, so that the surface adhering to the roadway was generally 24½ in. by 18 in., or about 2¾ square feet superficial. The traction power, in comparison with that of ordinary rigid wheels, is thus clear at a glance."

In July, 1871, another road steamer was made to draw two large omnibuses, filled with passengers, from Lincoln to Grantham, and the trip was in every way satisfactory. This engine and the omnibuses were made for Greece, and after their arrival at their destination, in 1872, they were again put to the test; the conveyances were loaded with seventy passengers, the engine hauling this load up an incline of one in seventeen at the rate of three miles an hour, and on the level road at eight miles an hour.

Mr. Thompson's "pot" boiler as made by Messrs. Robey and Co. for their road steamers is shewn in section by Fig. 74. "The boiler consists of a vertical steel shell, ⅜ of an inch thick, 2 ft. 9½ in. diameter and 7 ft. 9½ in. high. Within this is placed a fire-box 2 ft. 3in. diameter. Within the fire-box 's suspended the copper 'pot,' 2 ft. inside diameter, con-

Fig. 74

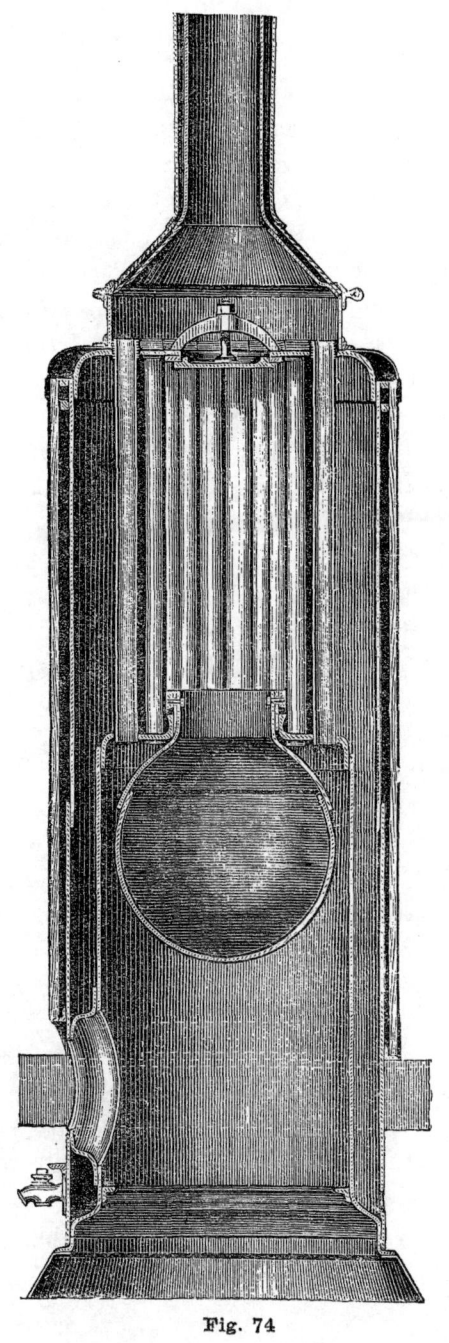

nected with the fire-box in the following manner : The 'pot' has a straight neck, as shewn, 4 in. long and 9½ in. diameter. The fire-box has a very similar neck into which the 'pot' fits easily. This neck is lightly bored out. On the neck of the 'pot' is secured a brass ring, outside which goes a second brass ring. The 'pot' is put into place from the inside, and the second ring is then secured to it by a number of half-inch bolts. The lower edge of the second ring and the upper edge of the fire-box ring are turned, and a ring of indiarubber is interposed between the two, and kept in place by making the surfaces slightly conical. The upper portion of the boiler is traversed by tubes 3 ft. 4 in. long and 2¼ in. in diameter.

We may here mention that Messrs. Robey and Co. were successful makers of road locomotives many years before Mr. Thompson's engines were built by them, but these early engines were chiefly designed for agricultural purposes, this being still an important branch of their large business.

RANSOMES. — The manufacture of Thompson's road steamers was undertaken by Messrs. Ransomes, Sims, and Jefferies in 1871. This renowned firm was established in 1789. The year before last was the centenary year of the Orwell Works. In passing, we may here briefly note the origin of the farm locomotive. The agricultural locomotive was born in the Eastern Counties. Messrs. Ransome, in Suffolk, and Messrs. Burrell, in Norfolk, were the first two firms to commence the regular manufacture of road locomotives in England. Several other firms in the early days made a great display, but their lights were soon extinguished and their names are now well nigh forgotten ; but the above-mentioned well-known firms have maintained a position in the front rank in this industry, and to-day are classed amongst the successful makers of locomotives for traversing common roads. At the Royal Agricultural Show at Bristol in 1842, Messrs. Ransome exhibited their first farmers' road locomotive, the engine and boiler were placed on a four-wheel carriage arranged to be self-moving, the power of the engine being transmitted to one of the driving wheels by means of a pitch chain. The Judges of the Show among other remarks said :—" The engine travelled along the road at the rate of four to six miles an hour, and was guided and manœuvred so as to fix it in any particular spot with ease, it turned also in a small compass." *The above was the first traction engine ever seen at an agricultural show.* Again at the Royal Show at Norwich, in 1849, Messrs. Ransome exhibited a neat little traction engine mounted upon springs with the two cylinders placed beneath the smoke-box. This engine was capable of running at the rate of five to six miles an hour, but the farmers were prejudiced against self-moving engines in those days. Not only were Ransomes the pioneers of steam locomotion on common roads; but they likewise made the first self-moving ploughing engine for Messrs. Fowler, which was shown at the Royal Show at Salisbury in 1857. This patriarchal engine was exhibited in the museum of old and new machinery, which constituted an interesting feature of the ever-to-be-remembered and disastrous show of the Royal Agricultural Society, at Kilburn, in 1879. From the foregoing remarks giving the barest sketch of the origin of the agricultural locomotive, it will be admitted that Messrs. Ransomes' firm was a most likely one to be entrusted with the manufacture of the high-speed engines for Mr. Thompson. These road steamers " had scarcely been tried and found successful when the Indian Government, with praiseworthy alacrity, purchased a small one of Mr. Thompson, with a view of testing its adaptability to passenger and goods traffic in that country." This engine was placed under the superintendence of Mr. Crompton, M.Inst.M.E., and although this engine was not a success, yet it proved that the rubber tyres were not affected by the climate, and on the whole Mr. Crompton's experiments were so satisfactory to the Director-General of the P. O. in India, that at his recommendation in August, 1870, Mr. Crompton was sent to England to superintend the construction of four larger engines. These were built by Messrs. Ransomes, Sims and Head, of Ipswich. One of the stipulations of the contract between Mr. Thompson and the Indian Government, when the former tendered for the supply of four high-speed road steamers, was that one engine at least should, as a test, travel a distance of several hundred miles, drawing behind it a load. The object of this was to learn as much as possible of their behaviour when put to severe continuous work. With this view, the first completed engine, the "Chenab," was sent by road from Ipswich to the Royal Show at Wolverhampton, 1871. The results of the trial, though satisfactory as far as the engine proper and the rubber tyres were concerned, were vitiated by the failure of the boiler, which could not be kept steam tight and otherwise gave endless trouble. At the Royal Show at Wolverhampton, after the various traction engines had been tested by running over a hilly course marked out on the Barnhurst farm, it was decided to further test the capabilities of the engines by a run of considerable length upon the high road, Instructions were given on 6th July to the exhibitors to make the journey from the show yard to Stafford, a distance of 16 miles. In addition to the five engines, the property of the exhibitors, which made this journey to Stafford, there was another Thompson engine, manufactured by Messrs. Ransome, which also performed the journey. This was the

"Chenab," which towed behind it an omnibus of a very unusual appearance: it was carried on a single pair of india-rubber tyred wheels placed in the centre. Its horizontal position was preserved by means of a long neck, containing at its outer end a nut, through which passed the vertical draft-pin of the engine. This pin was a screw, and by turning it the nut could be raised or lowered, so as to level the omnibus. The draw bar of the omnibus was elastic. Not only was there accommodation within the omnibus for passengers, but on its roof, there were four rows of seats for passengers, protected by a canopy. The sensation afforded by riding in the omnibus, although very peculiar, was by no means unpleasant. The "Chenab" was in charge of Mr. Crompton, who very kindly placed it at the disposal of the judges to accompany the competition engines on the run to Stafford. As much as eight miles an hour was accomplished by the "Chenab" on some parts of the road; but the engine was suffering the whole time from a deficiency of steam.* There is an amusing report of this run to Stafford and back in *The Engineer*, 14th July, 1871; from which we gather that the boiler was dosed occasionally by Mr. Crompton with red lead and oatmeal to stop the leaks, the engine had to be continually stopped to raise steam, and the Thompson's patent "pot" boiler was in many respects a failure. During the return journey from Stafford the copper blast nozzle was melted off, totally disabling the engine, which had to be put up for the night about five miles from Wolverhampton. Mr. Crompton retired for the night *in* the omnibus, and the men *on* the omnibus. *The Engineer's* correspondent decided to walk to Wolverhampton, and as he tramped along his dark journey he filled up his time in composing complimentary sentences respecting the ill-behaved 'pot' boiler. On the comple-

* Report of the Trials of Traction Engines at Wolverhampton, 1871.

tion of the second Indian Government engine, the 'Ravee,' the Field boiler of which gave excellent results when put through a series of trials with wood fuel on the Ipswich race-course, Mr. Thompson, before giving his final consent to the adoption of the Field-type of boiler in the remaining three engines, desired Mr. Crompton to put it through as exhaustive a trial as that undergone by the "Chenab," in order that the weak points might be found out (if any) of the Field system, and furthermore note the effect of a continuous high speed on the engine and the rubber tyres, this high speed having been unattainable with the defective boiler of the "Chenab." Three of the engines, according to the order, were to have been fitted with the "pot" boiler and one with the Field boiler; but all the four were fitted with the Field boiler. The "Ravee" made the celebrated double journey between Ipswich and Edinburgh in October, 1871. The following particulars are recorded of the trip. The total weight of the engine and omnibus was about 19 tons.

	Ipswich to Edinburgh.	Edinburgh to Ipswich.
Total fuel consumed in lbs.	29,148	25,312
Total water evaporated in lbs.	137,850	144,300
Total distance in miles	422·5	424
Time actually travelling	77 h. 26 m.	61 h. 13 m.
Engine standing under steam	72 h. 23 m.	42 h. 3 m.
Average speed in miles per hour	5·45	6·9
Gross load in tons	19	20
Mile tons	8027·5	8480
Lbs. of coal per ton per mile	3·631	2·98
Lbs of water per ton per mile	17·17	17·01
Lbs of water evaporated per lb. of coal	4·729	5·7

"By the return journey the men had ample experience in working the engine, and the 425 miles took nine days, giving 47 miles as the average distance per day, and a speed of

RANSOMES' INDIAN ROAD STEAMER AND TENDER.

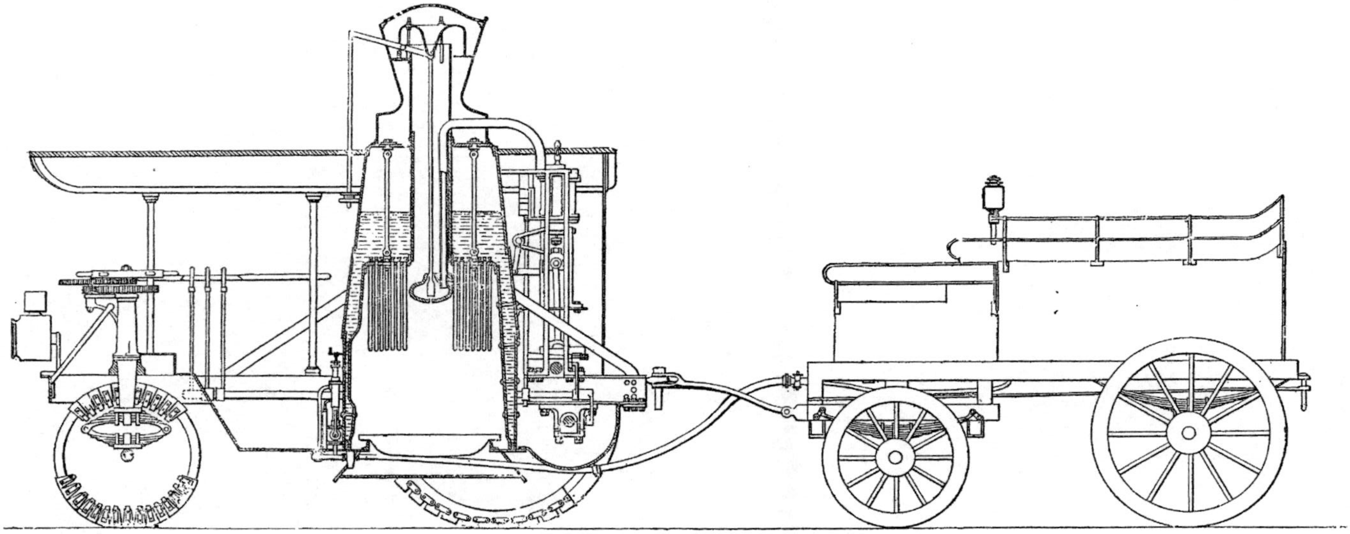

Fig. 75.

about seven miles per hour for the time actually running; but on the last day the average speed was about ten miles an hour, whilst occasionally a speed of from 15 to 20 miles per hour was maintained for short distances."*

Fig. 75 gives a drawing of the "Ravee," which shows its construction clearly. The cylinders were 8 in. diameter, and 10 in. stroke; geared either three and three-quarters to one, or twelve to one, on to driving wheels 6 ft. diameter. The engine made 150 revolutions per minute, which gave about ten miles per hour for the fast speed, and three miles per hour for the slow speed. The boiler was of the vertical Field system as shewn in section, the grate area was 11 sq. ft., and the heating surface was 177 sq ft. The blast nozzle had an adjustable cone, that the opening might be varied to suit either wood or coal. All the road wheels had Thompson's indiarubber tyres, with linked shoes; the leading wheels had supplementary elliptical springs, so that the front end adapted itself easily to the inequality of the road. The two-wheeled omnibus for carrying 65 passengers we have already described. These Indian Government locomotives had a large tank and stowage for wood so that they might run 15 miles without stopping. They were built for carrying mails and passengers between two stations in the Punjâb, about 70 miles apart. In 1879, Mr. Crompton read a most valuable paper before the Institute of Mechanical Engineers, on the working of these traction engines in India. Mr. Crompton says:—" several experiments with trains of wagons drawn by traction engines have been made in India, but the writer only describes the one which was carried out under his superintendence. The conditions of carriage which rule in our great Indian Empire are difficult and peculiar. Owing to the fact that agricultural produce forms the greater part of the freight carried, it follows

* "Rise and Progress of Steam Locomotion on Common Roads." Head, 1873.

that all channels of communication are fully worked for a brief period following harvest time, while for the rest of the year the capital employed lies idle and unremunerative. There is no legal limitation to speed in India. The Grand Trunk road is laid out with easy and regular gradients, its metalled surface is smooth and well formed and consolidated. Hence it was not unreasonable to expect that, if higher speed traction engines could be made successful anywhere, they would be so under such favourable conditions." *

In many points these expectations were fully realized by Mr. Crompton. A high speed was obtained without excessive wear and tear ; very regular running and timing were observed ; a large number of passengers were conveyed cheaply and safely, and goods were carried at a cheaper rate than by any land carriage other than rail. We are unable to give any particulars of the trials carried out in India as our space is limited. Those who wish to follow the subject further should procure Mr. Crompton's valuable paper. In concluding this notice of Messrs. Ransomes' Road Steamers, we ought to mention the fact that Mr. Crompton was obliged to reduce the weight of the engines from 14 tons to 9 tons, because the former weight was too much for the timber bridges of the Trunk Road ; and, moreover, the excessive weight of the engines ruined the rubber tyres. In India, the separate tenders on four wheels carried 500 gallons of water and the wood fuel, thus reducing the total weight of the engines by 5 tons. Fig. 75 shows the road steamer and separate tender.† Mr. Crompton, in speaking of the "Chenab," said :—" Once a driving wheel came off, twice her train became uncoupled, and finally her British driver got drunk and drove the engine off the road, with the result of turning her nearly

---

* " On the Working of Traction Engines in India. Crompton, 1879."
† Kindly supplied by the Proprietors of *Engineering*.

upside down. It was a tedious job to repair her after this last accident, but after the repair, *she worked for nine weeks, running night and day, covering nearly* 2,000 *miles without coming into the repair shop."* Just a line about the native drivers of whom Mr. Muirhead spoke so favourably. Mr. Crompton says :—" Some would say that natives could not be trusted with an engine that had to be steered, and with the lives of passengers. All he could say was (and we wish English traction engine drivers to note his remarks) that when he had European drivers he had narrowly escaped having two terrible accidents, the engines being very nearly driven over the parapets of a bridge ; that was entirely owing

Fig. 76.

to the great curse of drink. The native Mohammedan drivers do not drink ; they drove well and steadily and could be thoroughly depended upon." Messrs. Ransome made several road steamers of the same type as the one tested at the Wolverhampton Show, 1871, called the " Sutherland," but as these engines were constructed entirely for farm purposes we need not describe them. Messrs. Ransomes, Sims, and Jefferies manufacture traction engines and road locomotives, both simple and compound, in which are embodied all recent improvements.

200   MODERN PERIOD.

Fig. 76 shews Messrs. Ransomes' agricultural locomotive a large number have been supplied to customers in England and abroad, which are giving the greatest satisfaction. They are very simple in construction, and easy to manage.

BURRELL.—Messrs. Chas. Burrell and Sons, of Thetford, Norfolk, were one of the earliest makers of road locomotives; they commenced the manufacture of traction engines fitted with Boydell's endless railway wheels in 1856, and from that early period down to the present time, they have occupied a leading position in this branch of engineering. We have for many years closely watched the progress made by Messrs. Burrell in the design and construction of all types of road engines, and can fully endorse the statement made by *The Engineer* some time ago, who said : " Messrs. Burrell exhibited much fertility of invention in varying the design of their road locomotives for different classes of work." And as a proof of the soundness of their construction and workmanship, a few years ago the writer saw a traction engine of their make under repair after being in constant service for 22 years, and this engine was expected to do duty for some years to come after the repairs were completed. Messrs. Burrell commenced the manufacture of Thompson's road steamers early in 1871, and previous to the date of the Royal Show they had delivered four of these engines to customers, three of the first being sent to Turkey, all of which were of the same size and design as the one exhibited and tested at Wolverhampton, shewn by Fig. 77. This road steamer had two cylinders, each 6 inches diameter, and 10 inches stroke. The boiler was of the "pot" description, the two driving wheels were 5 feet diameter outside the rubber tyres, which were surrounded by a chain of steel shoes, the single steering wheel was

3 feet diameter; the steering wheel fork spindle passed upwards through a boss in a strong dome shaped casting, and was worked by a worm wheel keyed on it, driven by a worm on the horizontal steering handle. The cylinders were

Fig. 77.

inverted, and gave motion to a two-throw wrought iron crank-shaft, containing between the throws the four eccentrics for the link motion, these eccentrics were forged solid with the crankshaft. "At each end of the shaft there was a sliding pinion to gear into a spur wheel, bolted on to the inner side of the driving wheels. These pinions, when in gear, gave the quick speed for travelling. The crankshaft, by means of a spur pinion and wheel, drove the second motion shaft which extended across the engine, and carried at its two ends two pinions, which could be slid into gear for the slow speed with the spur gear of the driving wheels. This arrangement of gearing is illustrated by Fig. 72, given under Messrs. Robey's name. The sliding pinions on the crankshaft and the countershaft were controlled by levers, fixed on the ends of shafts which extended lengthways of the engine: those for the quick speed pinions were solid, and lay within the shafts for the slow speed. These latter shafts were hollow. The four handles were conveniently placed two on each side of the driver, and there was an ingenious arrangement of stops by which the driver was secured against inadvertently endeavouring to put the one speed into gear before the other was taken out. It will be seen that this disposition of handles sufficed for throwing out one wheel when turning sharp curves, as well as for putting in and out the quick and slow gear."* The whole of the working parts were enclosed in a sheet iron case. The boiler was intended to work at 130 lb. steam pressure, and the Judges of the Royal Agricultural Society said that the steaming of the boiler during the trial was very irregular.

During the earlier part of 1871 Messrs. Burrell and Sons wisely discarded the use of the 'pot' boiler, which gave so much trouble, and adopted the locomotive form of boiler. The design of the engine was entirely rearranged to suit the altered type of boiler. Fig. 78 shows this engine, the first of which was made for the Turkish Government, and it was pronounced by *The Engineer* "as one of the best

* Report of the Judges on the traction engine trials at Wolverhampton.

designed road locomotives up to that time constructed." Mr. John Head, in his paper read before the Institute of Civil Engineers,† also said that this engine " was an excellent specimen of careful design and workmanship." The horizontal engine was placed on the top of the locomotive boiler, the cylinders being fixed nearest the fire-box end of the boiler. The driving axle, coal bunker, feed water tank, and boiler

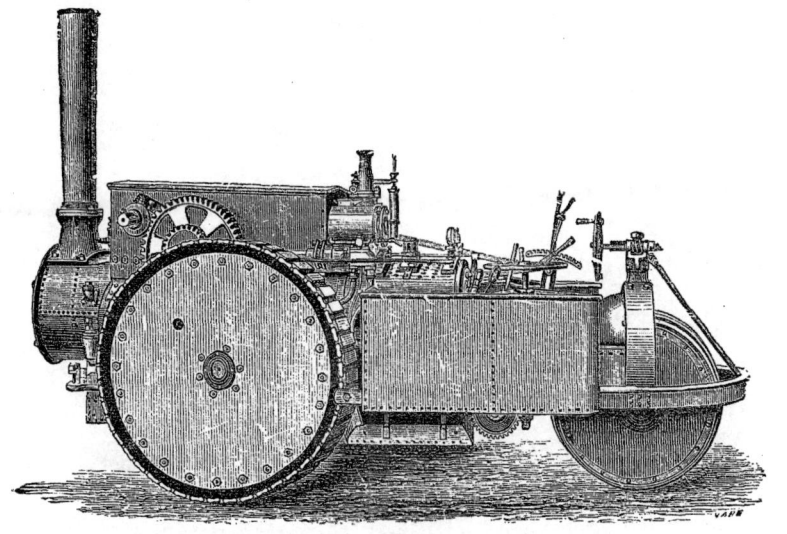

Fig. 78.

were all mounted upon a wrought-iron frame, as shown and carried by the four rubber-tyred travelling wheels. The two driving wheels were 6 feet diameter, the axle being situated about midway of the boiler barrel; the two leading wheels were 4 ft. diameter and placed close together. A single vertical pin rose from the front axle and was surrounded by a steel helical spring; the steering was effected by a worm and worm-wheel in the usual manner. By the illustration, Fig. 78, it will be seen that the fire-box of the boiler travelled fore-

† Rise and Progress of Steam Locomotion on Common Roads.

most, the fire-hole being at the side of the fire-box nearest the reader. An arrangement consisting of worm and wheel, pinion and rack, seen in the engraving, was fixed to the frame for raising or lowering the fire-box of the boiler when ascending or descending inclines. The two steam cylinders were each $7\frac{1}{4}$ in. diameter and 10 in. stroke. The engine had two speeds of steel gearing, the main spur gearing from the countershaft to the axle was provided each side of the engine. The gearing ratios were proportioned for running at four and eight miles an hour. It will be noticed that the greater part of the working details of the engine was neatly boxed

Fig. 79.

in. This engine was intended by Messrs. Burrell to have competed in the traction-engine trials of the Wolverhampton Show, but it was not finished in time, it was however exhibited at the show proper and attracted a great deal of attention. An engine of the same improved design was made by Messrs. Burrell for working passenger service in Greece, in connection with a large well-constructed omnibus intended to carry 50 persons as shown by Fig. 79. This road locomotive, it will be seen, is similar in design to the one we have just described and illustrated, except that the

BURRELL'S ROAD LOCOMOTIVE.

Fig. 80.

chimney was fitted with an American spark catcher, and an awning was erected over the engine driver and steersman. The omnibus was mounted on four wheels, unlike the ones used by Messrs. Ransome for service in India, because it was thought that some objectionable features were avoided by using the old and approved type of 'bus.* There was accommodation in the omnibus for passengers, and on the roof likewise, which was protected by a canopy. Another engine of similar design was sent to Russia. Fig. 80 shews

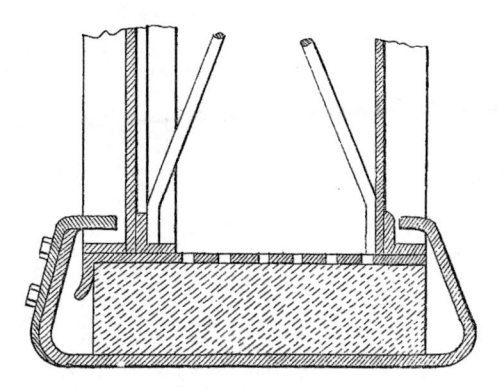

Fig. 81.

this engine, from which it will be seen that the two cylinders were placed at the chimney end, each 6 in. diameter and 10 in. stroke. The driving wheels were 5 ft. diameter, fitted with rubber tyres and Burrell's improved chain-armour or protecting shoes, shewn by Figs. 81 and 82—the leading wheel was 3 ft. 6 in. diameter. The travelling speeds were two and five miles an hour. The working steam pressure was 130 lbs. per square inch.

Figs. 81 and 82 shew Messrs. Thompson and Burrell's

* It would appear from Mr. Crompton's Paper, read before the Inst. of Mechanical Engineers, in 1879, on "Rubber-Tyred Traction Engines," that the two-wheel 'busses had been replaced by four-wheel ones of a smaller type.

improved shoes. It will be seen that the steel plates touched each other outside the tyre, and were tapered at the ends, each shoe being turned down so as to clip the angle iron ring on both sides, as shewn in the large scale section, Fig. 81. These shoes were a great improvement on Mr. Thompson's original ones, and answered well in practice.

In September, 1871, some interesting trials were carried out by Messrs. Burrell and Sons, at Thetford, with two of the passenger road locomotives we have illustrated and described. The engines tried were the twelve horse power for Crete in Turkey, and the eight horse power for Russia, and the large

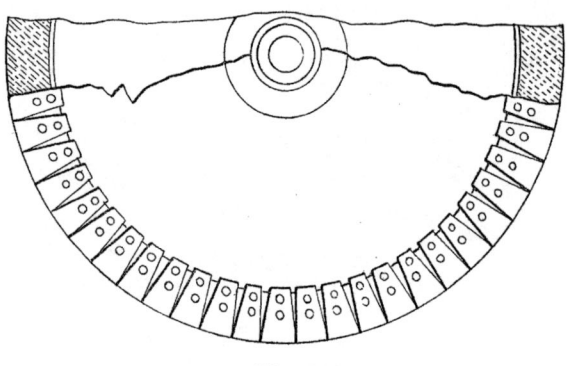

Fig. 82.

omnibus already illustrated. The trials were witnessed by Colonel Risa Bey, and Colonel Mehemed Bey, of the Turkish Service, and others. The large road engine, weighing 10½ tons, had the following load coupled up to it, two waggons loaded with pig iron, a portable engine, and a coprolite mill on wheels; the whole, weighing nearly 37 tons gross, the engine drew up and down the streets of the town at 5 miles an hour. In one place an incline of 1 in 18 was mounted without the slightest sign of any slipping; the engine would have drawn another 5 tons easily. The train was accompanied by an 8 horse power road locomotive for Russia. In the evening the 12 horse power traction engine was brought out again and attached to the omnibus. Men and boys swarmed over the 'bus like bees, the engine drew the load up hill and down, in the neighbourhood of Thetford, at the rate of 9 miles an hour, and in some places attained the speed of 12 miles an hour. The boiler was constructed for a working pressure of 150 lbs. per square inch.

The eight-horse power engine was then attached to the bus, and ran back to St. Nicholas Works faster than horses could go in regular service. After witnessing these trials the writer in *The Engineer* said :—" No difficulty whatever exists in applying steam on common roads to the purposes of passenger traffic." Messrs. Burrell and Sons made a considerable number of these well-designed road locomotives for quick speed travelling in foreign countries, and a goodly number were made for farm purposes in Great Britain, in which case, the cylinders were placed near the chimney, like Fig. 80, and the crankshaft with the clutch forks and the fly wheel for driving fixed machinery were placed at the fire-box end within reach of the driver. We need hardly say that the reputation of the Thetford firm is more than maintained by the well proportioned, and very efficient traction engines turned out by them at the present time. Their two latest patented improvements consist, first, of a method of mounting spur-geared road locomotives on springs, and second, of an arrangement of cylinder on the compound principle whereby a single throw crankshaft may be used.

Fig. 83 shews their road locomotive fitted with the spring arrangement and all their latest ideas.

For a long time Messrs. Burrell have kept the important problem of mounting traction engines on springs constantly before them, and more than thirty years ago their chain traction engines were mounted upon spiral springs fitted into the axle boxes, many of these engines are still in use. It is well known that the mounting of chain road locomotives upon springs is a very simple matter, but the application of springs to a spur-geared traction engine is a more difficult one, hence some of the plans hitherto introduced have been much too complicated to be of real service. It is not only necessary to arrange the details so that the up and down motion of the main axle, when passing over rough roads, shall not affect the pitch circles of the spur-

Fig. 83.

driving gear, or interfere with the proper action of the compensating apparatus ; but it is equally important that the springs should be free to act, without being influenced by the transmission of the power through the gearing. The patent spring arrangement was introduced by Messrs. Burrell at the Newcastle-on-Tyne Show of the R.A.S.E., 1887, and has been fitted to 60 engines during the last three years, and given great satisfaction. The range of the springs is sufficient to allow

BURRELL'S SPRING MOUNTED TRACTION ENGINE.

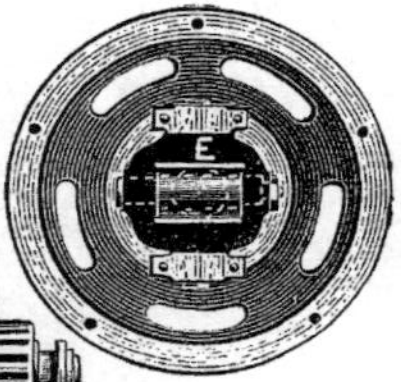

Fig. 86.

Fig. 84.

Fig. 85.

the engine to be driven at high speeds over ordinary highways. We are informed that, as the speed of the engine is increased, the vibration appears to be reduced, and the engine runs as smoothly as a carriage. Fig. 84 is a transverse section. Fig. 85 is a part longitudinal section of the Patent-Geared Traction Engine, mounted upon springs, while Fig. 86 is an end view of the gimbal or universal joint. These illustrations are so clear that very little description is needed.

In Fig. 84, A is the crankshaft, from which motion is communicated to the countershaft B by means of pinions gearing into the spur-wheel C. The spur-wheel C revolves upon a fixed steel tube D, and is connected to the countershaft B by means of the universal joint E. The other end of the countershaft is carried in the bearing F, free to move up and down in the box G The bearing F is connected by the link H to the axle box J (which is free to slide up and down in the guides S secured to the horn plates X) so that they both rise and fall together as the engine rides up and down upon the springs L. The pinion M and the spur-wheel N are prevented from altering the distance between their centres, and so getting out of gear by the link H, and have also sufficient clearance at the sides of the teeth to allow of any sideway motion caused by the action of the engine on the springs. The axle-box K is also free to slide up and down in the guides S, its motion in no way altering the distance between the centres of the gearing. As shown in Fig. 84 the steel tube D is carried quite across the engine and bolted to the box G, which is also of steel and fastened to the hornplate X, thus forming a solid bearing for the spur-wheel C and a substantial stay to the horn-plates X, and effectually preventing them being twisted or buckled by the strains thrown upon them. The volute springs LL can readily be adjusted by tightening up the nuts under them, or can be

P

quickly changed if necessary without taking any other parts of the engine to pieces. The front of the engine is carried upon one spring which is fitted in the fore-carriage casting, and can be easily adjusted by removing a cap in the smoke-box.

We will now describe Messrs. Burrells' single crank system of compound cylinders. Fig. 87 shows the front part of a road locomotive fitted with the compound cylinders, which

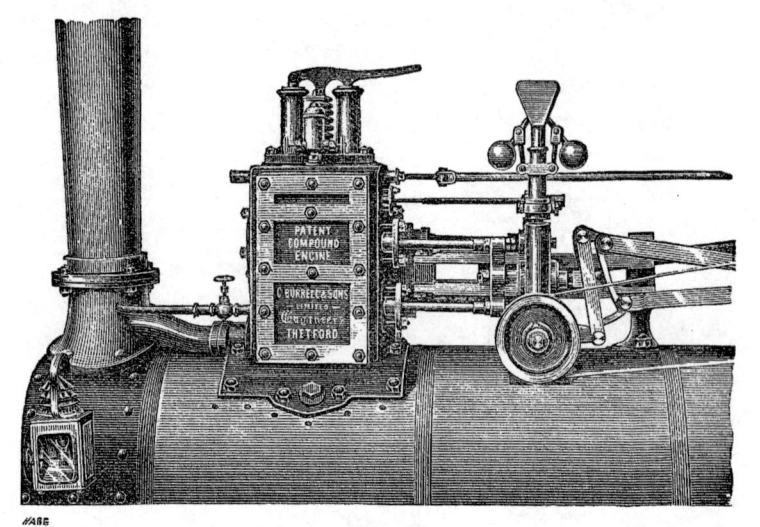

Fig. 87

was introduced at the Windsor Show of the R.A.S.E., 1889. The high pressure cylinder is placed diagonally over the low pressure cylinder, and the piston rods of both are connected to one long crosshead, which is equal in length to the distance between the two centres of the cylinders. This crosshead has inside flanges bearing against the edges of the motion bars, and thus presents a large surface for wear, and takes up any cross strain. The steam after leaving the smaller cylinder passes directly into the low pressure cylinder, all clearances being filled up as closely as possible. The slide valves of both the high and low pressure cylinders are coupled together and driven by one link motion, so that beyond having two extra glands to pack, there are no more parts requiring attention than in the ordinary high pressure engine with one cylinder. Both pistons are always moving together at the same velocity, so that the flow of the steam through the cylinders is uniform, and immense power is developed in starting a load. Engines thus fitted are as easily handled as single cylinder engines. By this neat arrangement the full benefit of the compound system is secured, thus effecting a saving of 30 per cent. in fuel, and without any complication. Not only is there a saving in fuel, but the saving of wear and tear to the boiler also is most important, the work upon the boiler being lighter, and consequently the heat in the fire-box much less. This is owing to the low pressure at which the exhaust steam is discharged up the funnel. Compound engines work more noiselessly on this account, and are less liable to frighten horses upon the roads than single cylinder engines. The value of both these improvements is greatly enhanced owing to the fact that they can be applied to existing engines.

AVELING.—Messrs. Aveling and Porter, of Rochester, have during the last thirty-two years devoted themselves almost exclusively to the manufacture of traction engines ploughing engines, and steam road rollers, and during this period their name has been associated with numerous valuable patents relating to this branch of engineering. In 1858, the late Mr Thomas Aveling designed and patented an arrangement for making portable engines self-moving, he applied a driving chain for communicating the power from the crankshaft to the axle, and in other ways modified the construction of Messrs. Clayton and Shuttleworth's engines, so that they would propel themselves from farm to farm, and haul a thrashing machine behind them. Judging by the testimonials given in Messrs. Aveling's catalogues of thirty years ago, many a portable engine was altered so as to do without horses, to the delight of the owners whom, we need scarcely add, after having one engine fitted with the locomotive gear and front steerage, lost no time in ordering the necessary castings for altering the rest. In some of the traction engines of these early times, a horse was placed in a pair of shafts for steering the engine along the road, but Mr. Aveling arranged a very simple steering gear, illustrated on a previous page in this book. In 1860, Messrs. Aveling shewed a chain traction engine at the Canterbury Show of the R.A.S.E., and at the London Exhibition, 1862. Referring to both these exhibits we quote the following from the firm's catalogue for 1863: "In the Exhibition of 1851 there was not one traction engine, and it was the general belief that none could be contrived to answer any commercial purpose. In the 1862 Exhibition nine of these engines were shown in the eastern annex. At the Royal Show at Canterbury, in 1860, Mr. Aveling exhibited a self-propelling engine—this was regarded with indifference by the officers of the society, and was catalogued with the miscellaneous articles." *The Engineer*, in 1862, said "Mr. Aveling's traction engine is the best, we think, that has yet been produced." In 1864 one of these road engines, named "El Buey," made for the Traction Engine Company of Buenos Ayres, hauled a load of 28 tons up Star Hill, at Rochester, which incline is 305 yards long, and rises one in twelve the whole distance; the engine also drew a train of 22 tons over soft ground in the locality of the Common, crossing a ditch 4½ ft. wide and 2 ft. deep.

We should be delighted to follow up the history of Messrs. Aveling and Porter's many improvements, but we will briefly refer to an engine exhibited at Oxford, in 1870, and then leave the agricultural types to notice the quick-speed road locomotives.

Up to 1870 all the Rochester firm's engines were of the chain type, but the little five-horse power engine, shewn at the Royal Show at Oxford, was driven entirely by gearing, and steered from the foot-plate. In this engine the ordinary cast-iron crankshaft brackets were dispensed with, the plates of the fire-box sides being carried upward and backward to serve the same purpose, and fitted with suitable bearings, while a neat cross casting served to render them perfectly steady.

About this time Mr. Thompson's road steamers were being made by Messrs. Burrell and others, fitted with rubber-tyred wheels, as previously described. We mentioned under Mr. Thompson's name that the indiarubber was not a suitable material for taking any driving strain, it should only be used as springs. Fig. 68 shews the creep of the rubber tyre. Now, Messrs. Aveling and Greig's patent rubber-tyre wheel of 1870 was introduced to rectify this defect.

Figs. 88 and 89 shew this wheel in section, from which it will be seen that the indiarubber is attached to the tyre of the wheel in segments, by a process patented by Messrs. Sterne and Co. The great advantage of this plan was, that if a segment got damaged it could easily and quickly be removed, and replaced by a spare segment at moderate cost. To avoid all possibility of slip in wet weather, and on clay soils, Mr. Aveling introduced the steel angle-iron crossbars, arranged so as to take the traction, without neutralising the benefit derived from the elastic action of the rubber. As the rubber yielded, affording the flexible broad bearing on the

214 MODERN PERIOD.

road, which was the secret of the success of the Thompson engine, the steel straps could slide between the intermediate guides. The rubber was thus protected; and from severe ex-

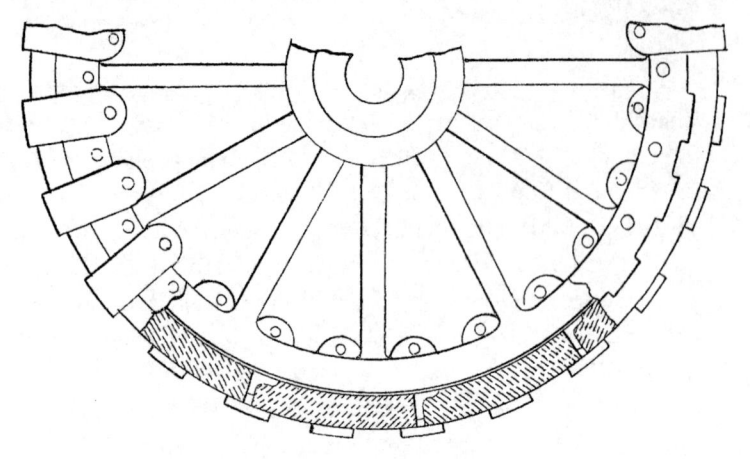

Fig. 88.

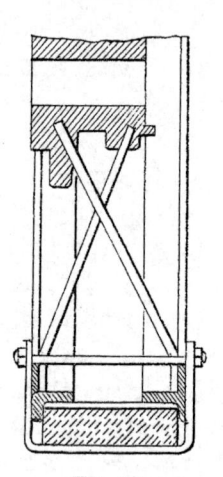

Fig. 89.

periments carried out by Messrs. Aveling, the wheels tyred in this manner proved a success.

In 1871, Messrs. Aveling and Porter commenced to build high-speed road engines, specially equipped for military purposes, called "Steam Sappers"—engines of this type have been purchased by the French, Italian, and the English Governments. One was tested at the Wolverhampton Show of the R.A.S.E., having a cylinder 7¾ in. diameter and 10 in. stroke. The gearing ratio was seventeen to one. This

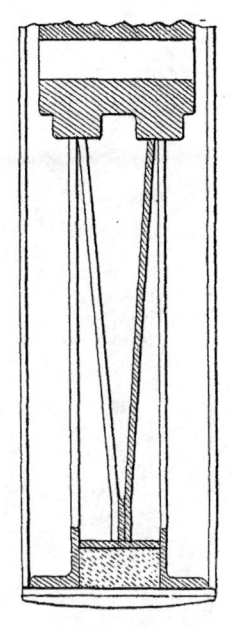

Fig. 90.

engine was fitted with Mr. W. Bridges Adams's spring driving wheels, which are represented by the following illustrations: Fig. 90, shews a section of the wheel. Fig. 91 shews a pair of these wheels very clearly — one complete, and the other with one of the angle-iron rings removed so as to show the indiarubber blocks between

BRIDGES ADAMS'S WHEEL

Fig. 91.

the inner and the outer tyres. The spokes of the wheel are riveted to a strong tee-iron ring which forms the inner tyre, while outside this there is a hoop of sufficient strength, stiffened by an angle-iron ring, placed near each of the outer edges of the tyre, and provided on its face with the usual diagonal strips for giving the wheel increased adhesion. Between the inner tyre of tee-iron and the outer ring just described are inserted blocks of indiarubber, which are kept in place sideways by the angle-iron rings. These inner and outer tyres are connected by a drag-link clearly shown by the engravings, which prevents any friction taking place on the rubber blocks. This wheel is simple and has proved durable, but the cost of the indiarubber was too great to compensate for the advantages gained. Messrs. Aveling have continued to build road engines suitable for running at a speed of five or six miles an hour with light loads, and slower when hauling heavy loads, as used by various Governments. These military engines or "steam sappers" have been of Messrs. Aveling's standard types, slightly modified to suit the requirements, and they have given the utmost satisfaction.

We must now describe the fine road locomotive introduced in 1878, one of which was shown at the Bristol Show of the R.A.S.E., and another at the Paris Exhibition of that year. One of the chief features of the engine was the arrangement of the gearing inside the bearings, and not on the overhanging ends of shafts outside the bearings. Everyone acquainted with the working of gearing subject to very heavy strains will know how to appreciate this improvement; the bearings are more fairly worn, and the gearing is maintained rigidly in truth. Another important gain in the arrangement patented by Mr. Aveling, is, that at the same time small pinions and large spur wheels are alike dispensed with, the employment of an intermediate shaft, securing the reduction of speed between the crankshaft and the driving wheels. The slow gear ratio being 26 to 1. This engine was fitted with Messrs. Aveling's patent crankshaft brackets, which are formed out of the side plates of the fire-box extended upwards and backwards, carrying the crankshaft, countershaft, and driving axle bearings in one plate.

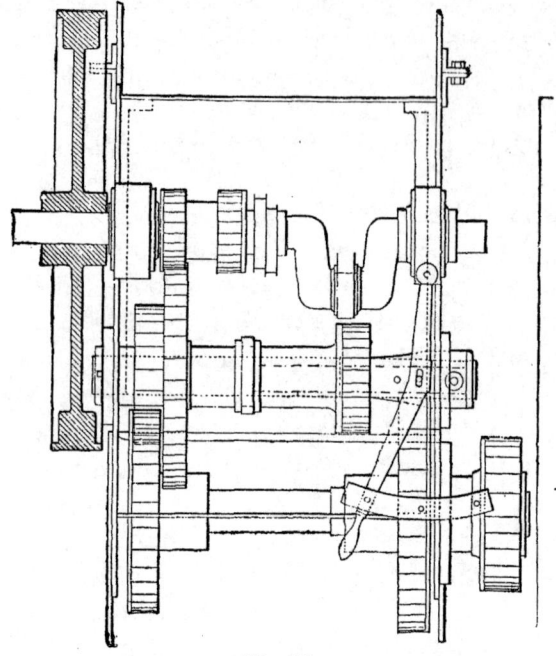

Fig. 92.

The driving wheels are 7 ft. in diameter, and 16 in. wide on face, and are constructed of a specially strong section of tee iron. The winding drum on the hind axle is capable of holding 100 yards of ¾ in. diameter wire rope, the compensating gear wheels are of large size. The boiler was intended for a working steam pressure of 150 lbs. per square inch, the fire-box stays being pitched four inches apart.

Fig. 92 gives a plan of the gearing, while Fig. 93 shews a transverse section through the hind part of the engine. From these views it will be seen that the whole

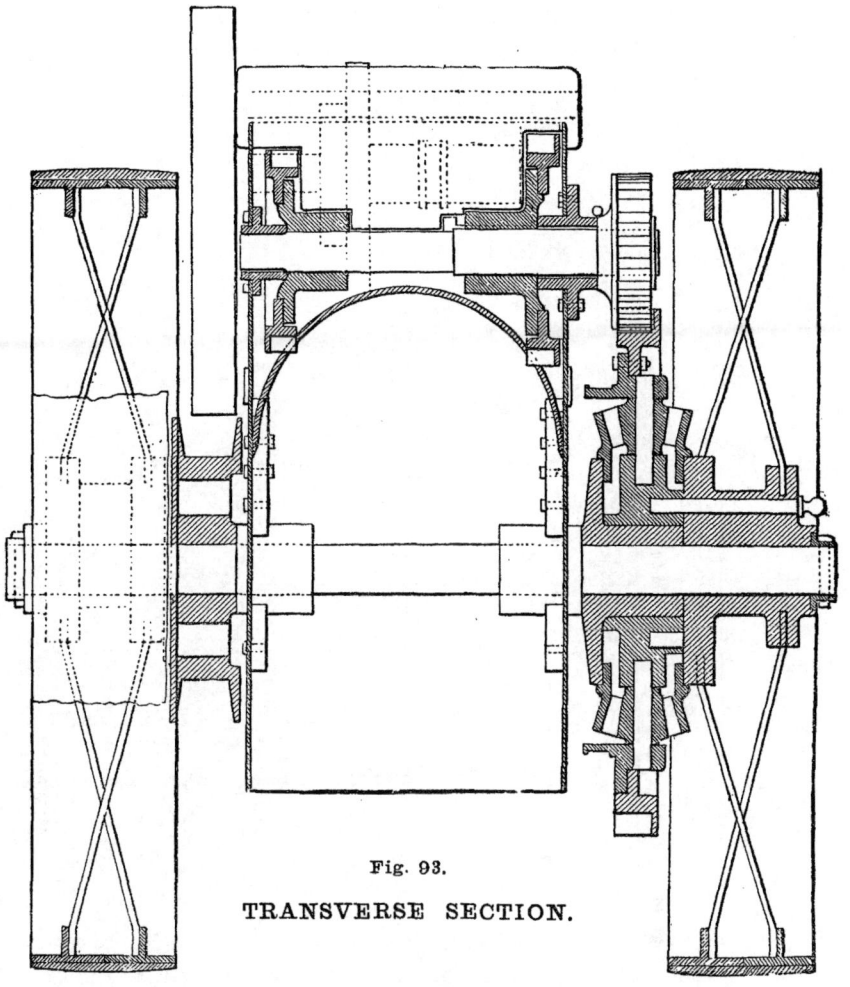

Fig. 93.
TRANSVERSE SECTION.

of the crankshaft and countershaft gearing is arranged to work between the wrought iron brackets, and the fly-wheel is

220   MODERN PERIOD.

fixed close to the crankshaft bearing. The pinions for the two speeds are keyed fast upon the crankshaft, instead of sliding on "feathers," the arrangement for altering the speed is shown in plan Fig. 92. The intermediate shaft is fixed, and serves as a stay to the side plates, and the sliding

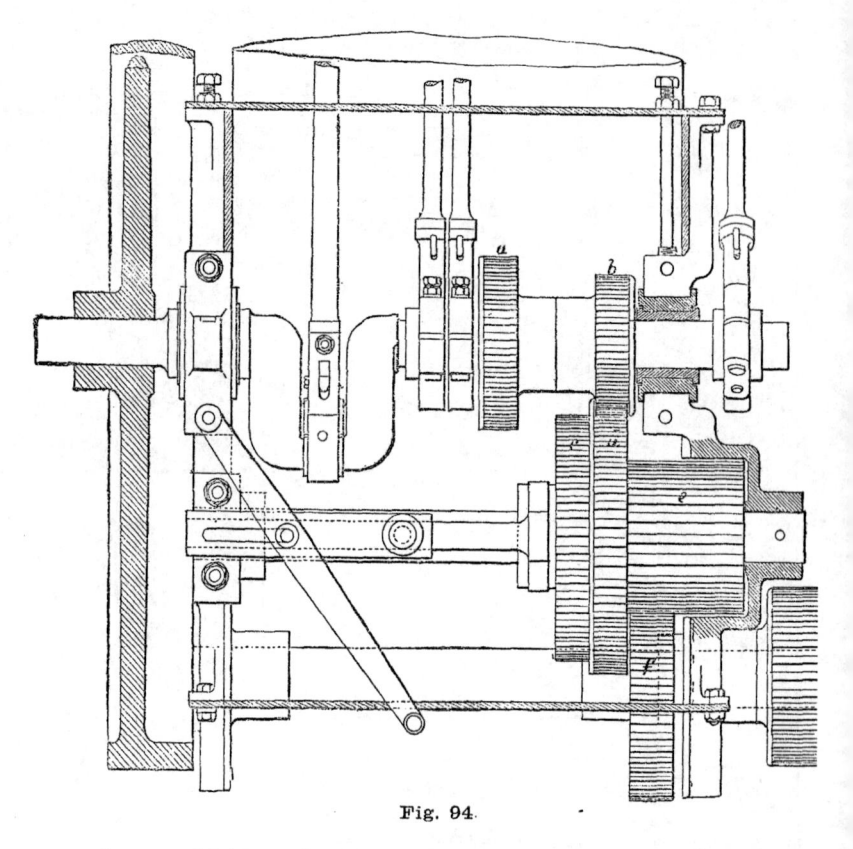

Fig. 94.

sleeve, which carries the spur wheel and the fast and slow speed pinions, revolves on it. The two crankshaft pinions are of the same size, and the intermediate spur wheel gears with one or the other, as required.

When the parts are in the position shewn in the plan, the engine is in its fast gear ; but if the sleeve be moved to the right, the other crankshaft pinion comes into gear ; also the right hand sleeve pinion gears into the right hand spur wheel upon the countershaft, and the engine is in slow speed. The engine is stronger, narrower, more compact in consequence, and the wear and tear most certainly reduced.

Without staying to refer to many details of interest, we pass on to notice another arrangement of gearing exhibited by Messrs. Aveling and Porter on one of their traction engines at the Smithfield Show, 1883. Fig. 94 shows this arrangement of inside gear, kindly supplied by the proprietors of *Engineering*, from which it will be seen that the crankshaft carries two pinions a and b keyed fast on it. The intermediate shaft is fixed, and on it turn the two wheels c and d and the long pinion e, the two wheels just named being cast together and bolted to the pinion. In the positions shown in the engraving the engine is in slow gear, the pinion b on the crankshaft engaging with the wheel d on the intermediate shaft. By means of the clutch and lever shown, however, the wheels and pinion $c\ d$ and e can be slid to the left along the intermediate shaft so as to make the wheel c gear into the pinion a, and thus get a faster motion, the pinion c being of such a length that it remains in gear with the wheel f.

Fig. 94 also shews Messrs. Aveling and Porter's cast-iron plate bracket for carrying the crankshaft, fixed intermediate shaft and countershaft bearing. It will be seen that these cheeks are secured to the outside of the fire-box side plates, and have flanges provided for bolting the transverse plates to them, making a sound and thoroughly mechanical job.

It is impossible to speak too highly of the well-designed and latest pattern road locomotive shewn by Fig. 95. The

AVELING AND PORTER'S ROAD LOCOMOTIVE, 1890.

Fig 95

ROAD LOCOMOTIVE AND DYNAMO, 1890.

Fig. 96.

cylinder is thoroughly steam jacketed, and placed as near the smoke-box as possible, The side plates of the fire-box shell are carried upwards and backward for taking the four bearings of the various shafts. The first and second motion gearing is placed inside the bearings as before. An extra tank is provided beneath the boiler barrel, connected by a pipe with the tank under the foot-plate. The working parts of the engine are neatly boxed in, and the fly-wheel is of the disc pattern so as not to frighten horses. The driving wheels, compensating gear and draw bar are all exceedingly strong, and intended for rough usage and continuous road haulage work. Messrs. Aveling and Porter's road locomotive and dynamo are shewn by Fig. 96—engines of this type were sent to Suakim for military purposes.

FOWLER.—Another celebrated and well-known firm engaged in the manufacture of road locomotives is Messrs. John Fowler and Co., of Leeds, who have during the last thirty years or more, made the construction and use of traction engines their special study, having in this lengthened period carried out the most extensive and exhaustive experiments quite regardless of expense.

In passing we may mention that Messrs. Fowler occupy a pre-eminent position as manufacturers of steam ploughing machinery, as their prize list will show. We have only space to name some of the leading improvements this eminent firm has effected in road locomotion. At the Wolverhampton Show of the R.A.S.E., in 1871, Messrs. Fowler exhibited two twelve horse-power road locomotives containing some interesting features, one of these engines was fitted with Aveling and Greig's wheels, with indiarubber tyres in segments, as described under Messrs. Aveling's name. (See Figs. 88 and 89.) One of these engines was mounted on three rigid wheels, the front wheel turned in a ring, supporting the weight of the forward part of the engine either by a number of balls or by rollers, the fore-carriage thus formed being steered by means of a worm, worm-wheel and pitched chain. This engine was provided with a single cylinder, completely steam jacketed and lagged, and combined with a spacious dome, the safest and simplest remedy against priming; steam was taken from the smoke-box end of the boiler, the driest part of the steam space. The fire-box shell side plates were carried up to support the shafts in the well-known manner introduced by Messrs. Aveling and Porter. A tank under the boiler barrel was provided, while all the moving parts were cased in to avoid frightening horses. In 1878, Messrs. Fowler introduced a simple arrangement of inside gearing—Aveling and Greig's patent. But the two chief points Messrs. Fowler have for years incessantly aimed at are, first: the perfecting of the compound cylinder system to road locomotives; and second: mounting the engines on an effective arrangement of springs. These two problems have been successfully solved. The peculiar advantages of both improvements are very well known and appreciated.

We now pass on to notice Messrs. John Fowler's recent type of compound road locomotive mounted upon springs, as illustrated by Fig. 97.

By the use of a low pressure cylinder working in conjunction with a high-pressure cylinder, the greatest possible amount of expansion is obtained, and almost all the power in the steam is used before it is discharged into the atmosphere, the pressure of the exhaust steam being only 8 lbs., which is just sufficient to create a draught necessary for the combustion of the fuel. These engines will do the same amount of work with 30 per cent. less fuel and water than the ordinary engines

FOWLER'S COMPOUND SPRING MOUNTED ROAD LOCOMOTIVE, 1890.

Fig. 97.

in every-day use. For example, where an ordinary engine consumes, say 6 cwt. of coal for a day's work, a compound engine will only require 4 cwt., and with the same quantity of water to start with, the compound engine will travel at least one-third greater distance than the ordinary high-pressure engine. In consequence of the steam passing from these engines into the atmosphere at the low pressure of 8 lbs., the noise caused by the exhaust, and all dangers from sparks are avoided, and the engine rendered much less objectionable to horses than the ordinary single-cylinder engine. The crankshaft and second-motion shaft are supported by the fire-box shell side plates, which are carried upwards, and strongly stayed by cross plates.

The injurious strain of the engine on the boiler is thus reduced to a minimum, and the weakening effects of bolting the brackets on the boiler avoided. The bearings are accurately fitted into recesses cut in these plates, and are also securely riveted to them. By this arrangement all the strain is taken by the plates themselves, the rivets merely hold the bearings in their places, and prevent any danger resulting from them giving way.

To combine the greatest possible strength with the requisite lightness, the boiler is entirely constructed of steel. The fire-box has a round instead of a flat top; and there is no danger from sediment accumulating on the top of the fire-box. All the longitudinal seams are double riveted, plates planed on the edges, and the flanging done by hydraulic pressure. The boilers are strongly stayed throughout, to work at a pressure of 150 lbs. per square inch, and tested to 250 lbs. per square inch, and are equal in strength and workmanship to the best locomotive boilers. The draw bar is so arranged that the whole of the pull is taken by the boiler side plates, and the tank relieved of all strain.

This engine shewn by Fig. 97 is mounted on springs. Messrs. Fowler refer to the spring arrangement as follows " The want of springs on traction engines, especially those used for running on hard and rough roads, has long been felt, and has proved a hindrance to their extended use. It has limited their speed, and increased the wear and tear. After experimenting for many years to find some means of allowing the shafts on which the gearing runs, to vary their relative positions without altering their relative distances apart, and thus to allow the play of the springs carrying the engine and boiler, we have now succeeded in doing this by firmly fastening two shafts parallel to each other, but allowing one of them to move round the other. By this means the pinion moves round the wheel it is driving, and allows the springs to move freely."

By the spring arrangement Messrs. Fowler claim these advantages : Rigid driving wheels; rigid gearing; the tractive power taken by the horn blocks and not by the springs; no extra wearing parts; the jar on the road reduced to a minimum; less wear and tear on the roads; little or no noise; no vibration (very essential when passing through towns); cost of repairs reduced at least one-half.

The following table gives the dimensions of the engine we illustrate, fitted with compound cylinders :—

COMPOUND ROAD LOCOMOTIVE.

Diameter of Cylinders	$6\frac{1}{2}$ in. and $11\frac{1}{4}$ in.
Stroke	12 in.
Diameter of Fly Wheel	4 ft. 6 in.
Width of Fly Wheel	6 in.
Diameter of Driving Wheels	7 ft.
Width of Driving Wheels	16 in.
Diameter of Leading Wheels	4 ft. 6 in.
Width of Leading Wheels	9 in.
Fast Speed on the Road	3 to 5 miles.
Slow Speed on the Road	$1\frac{1}{2}$ to 2 miles.
Revolutions of Engine per minute	150
Working Pressure per square inch	140 lbs.

HORNSBY'S COMPOUND ROAD LOCOMOTIVE, 1890.

Fig. 98.

HORNSBY.—Messrs. R. Hornsby and Sons, of Grantham, commenced the manufacture of traction engines about 1864. Their first road engines were made to the plans patented by Messrs. Hornsby, Bonnall and Astbury, dated July, 1863. This patent engine had two cylinders placed between side plates, and fixed beneath a locomotive multitubular boiler. The cylinders were arranged with trunks, or hollow piston-rods, passing through both ends of the cylinder covers. The small ends of the connecting rods were attached in the middle of the trunks, on the centre line of the pistons. No slide bars were required, and good long connecting rods could be adopted, and at the same time allow the cylinders to be pretty close to the crankshaft. The cylinders were placed beneath the smoke-box, and by this arrangement of engine, the cylinders, crankshaft, gearing, and countershaft could be nicely accommodated beneath a short boiler barrel. Carriages for the two above-named shafts were fixed to the deep side plates (railway locomotive fashion). Motion was communicated from the crankshaft to the countershaft by spur gearing; and from the countershaft to the driving axle by a pitch chain of novel construction. The main axle was placed across the fire-box front, beneath the fire-hole door, and mounted on volute springs in a similar manner to some of the modern plans. The working parts of the engine were all neatly boxed in, to protect the wearing surfaces from the dust. The engine was provided with a fly-wheel on the crankshaft, brake on the axle, and the engine presented a neat and trim appearance. One of this type was sent to Natal.

During later years Messrs. Hornsby have entered extensively into the manufacture of traction engines, for farm and other purposes, on the simple or the compound principle. We may briefly name the following good points relating to their narrow-gauge traction engines. The brackets are all bolted to wrought-iron plates, which are *riveted* to the boiler, avoiding the risk of leakage. The driving pinions are fitted on solid keys on the crankshaft, so that badly fitted keys are avoided. The engines are made unusually narrow in width by an improved arrangement of the gearing. We shall refer to Burrell and Edwards's clutch gear presently. The brake disc is cast with the left-hand road wheel, so that it cannot be rendered useless by the driving pins coming out. When the engines are used for driving machinery from the flywheel, the belt may be carried in either direction, this is, past the smoke-box, or past the tender, running quite clear.

Messrs. Hornsby and Sons make road locomotives either simple or compound, and by their courtesy we are enabled to show their most recent type, as per Fig. 98. The illustration represents their six-horse power nominal compound road locomotive. The cylinders are 5 in. and 8 in. diameter, and 12 in. stroke. The driving wheels are 6 ft. diameter and 14 in. on face, and the leading wheels are 4 ft. diameter and 9 in. wide. An engine of this size and type was recently fitted by Messrs. Hornsby with Messrs. Aveling and Porter's Patent Spring Wheels, which were a complete success. Our readers will remember that these wheels consist of an inner ring or tyre attached to the spokes, and suspended within an outer ring or tyre by springs suitable to resist either a tension or compression. The springs are held by worm pieces screwed into the coils, and carrying eye-bolts at each end, one bolt is attached to each spoke, and the other eye-bolt of each spring is attached to the outer tyre. When the wheels are used as drivers the springs are all under tension or compression. Fig. 98 shows an engine mounted on the wheels of the ordinary construction. Returning to the description of our illustration the slow gearing ratio is 22·6 to 1; the fast

MODERN PERIOD. 231

gearing ratio is 12·75 to 1. The engine is fitted with Messrs Burrell and Edwards's* patent clutch gear. Fig. 99 shows a side view of the gearing, and Fig. 100 gives a plan of the arrangement. These blocks are kindly supplied by the proprietors of *Engineering*.

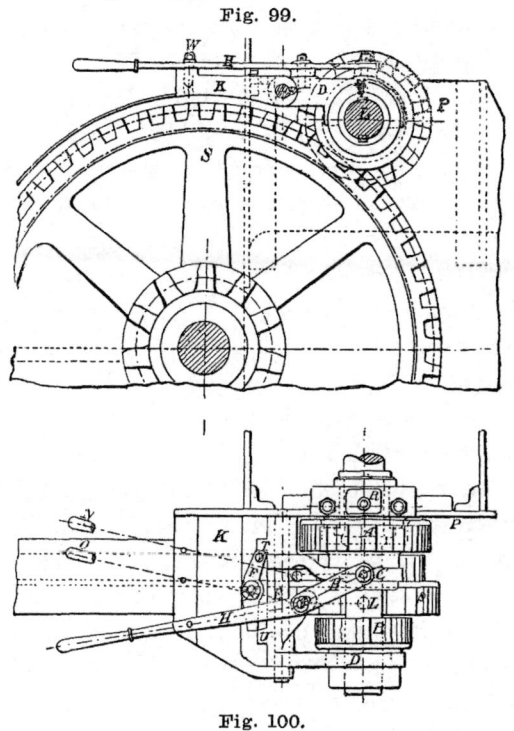

Fig. 99.

Fig. 100.

The following is a description :—" According to the plan the fast-speed pinion A on the crankshaft is placed inside and nearest the bearing; the slow-speed pinion B outside it. The two pinions are held laterally by forks C and D, which fit into grooves turned in their bosses, the other

* Mr. R. Edwards was for several years chief draughtsman for Messrs. Burrell and Sons, and he now occupies the position of engineer at Messrs. Hornsby's works.

ends of the forks sliding on a round bar E, which is attached to a bracket K, mounted on the side frame P. The forks just mentioned have projections T and U on their backs, as shewn, while a horizontal plate F swivelling on a pin X, and capable of being moved by a small handle V, is arranged so that its ends can engage with the projections T and U on the sliding forks C and D respectively, thus holding either of them stationary. A lever H, having one end working on a pin on the fork C, and the middle on a pin on the fork D, is used for sliding the pinions in and out of gear, the action being as follows:—With the parts in the position shown in the engravings, both pinions are out of gear, and the locking plate F is holding the fork C, and consequently the pinion A. On taking out the pin W the lever H can be moved to N, thus moving the fork D and its pinion B into gear, the fork C (being held stationary by F) the pin on it acting during this movement as the fulcrum to the lever H. If, however, on the other hand, it is desired to put the pinion A into gear, the locking-plate F must be pulled back by V, when the other end of it will engage U and hold the fork D and its pinion B stationary and out of gear. Under these circumstances the pin Y on the fork D acts as a fulcrum to the lever H which can be then moved to O, sliding the fork C and its pinion A into gear.

It will be noticed that with this arrangement both pinions cannot be placed into gear at the same time, as the plate F is so arranged as to lock one fast, while the other is free to be moved. The whole device is very neat and simple, and it has the advantage that only one lever is employed to shift the pinions.

It is very well known by traction engine experts, that the plan so often adopted by traction engine makers, of causing the slow gear pinion to slide inside the fast gear, in order to

MODERN PERIOD.

get both of them as near the bearing as possible when driving, has one great fault among others, viz., the fast gear pinion is made too large, giving too much difference in the gearing ratios between the slow gear and the fast gear consequently the fast speed gearing is much too fast to be of any practical use. Now, by the adoption of this patent gear this difficulty is obviated, the large pinion is placed nearest the bearing and can partly slide over it, while the small pinion is moved out of gear by sliding outwards, therefore these pinions can be made exactly the size found to be most useful in practice. Suppose, for instance, that 12·5 to 1 and 25·0 to 1, were the fast and slow speed ratios adopted by a firm using the faulty plan of one pinion sliding inside the other, it is well known that 15 to 1 and 25 to 1 would give far better results, and could be adopted by using the device we have just described. We may note here, that those makers who arrange some of the gearing between the bearings, and have two countershafts, can also adopt any proportions of pinions required. But Messrs. Burrell and Messrs. Hornsby urge that a second countershaft is of doubtful utility. We reserve our views for the present, and leave our readers to form their own opinion on this point.

The following table gives particulars of the gearing adopted by Messrs. Hornsby on their six-horse road locomotive.

	SLOW SPEED.	FAST SPEED.
Spur Pinion on the Crankshaft	13 teeth 1¾ in. pitch.	20 teeth 1¾ in. pitch.
Spur Wheel on the Countershaft	53 ,, 1¾ ,,	46 ,, 1¾ ,,
Main Spur Pinion	11 ,, 2¼ ,,	11 ,, 2¼ ,,
Main Spur Wheel	61 ,, 2¼ ,,	61 ,, 2¼ ,,

Referring to Messrs. Hornsby's road locomotive shown by Fig. 98, it will be noticed that the fore tank is placed beneath the boiler barrel and fills up the space between the saddle plate and the front axle. At the fire-box end room is provided for the steerage shaft and chains to run clear of the tank, a dotted line on the drawing shews the end of the tank. An opening is cut in the side of the tender, which renders access to the foot plate far easier than when the engine driver has to climb over the side. The draw-bar takes hold of the horn plates of the boiler as shown. By the arrangement of the weigh-bar shaft being placed close to the cylinder, long eccentric rods are obtained, which is a great advantage, a much better distribution of the steam is effected thereby. One bracket carries the four guide bars and the governors. Only one countershaft is used; Burrell and Edward's clutch gear, described and illustrated by Figs. 99 and 100, is applied. The engine presents a very neat appearance, the design having been very carefully worked out. Engines of this type are sure to be appreciated.

MACKENZIE.—A steam brougham of very neat appearance was made by Mr. H. Mackenzie, Scole, near Diss, in 1874. The two cylinders were each $3\frac{3}{4}$ in. diameter and $4\frac{1}{2}$ in. stroke. Power was transmitted by gearing from the crankshaft to the countershaft, and thence by pitch chain to the driving axle, the ratios being 6 to 1, and 13 to 1, to produce the two speeds. A "Field" type of vertical boiler, 2 ft. diameter and 4 ft. high, was used, intended for working at 135 lbs. pressure per square inch. The driving wheels were 4 ft. diameter. A single steering wheel was actuated from the inside of the carriage. Four passengers could be accommodated inside the vehicle.

PERKINS.—We have now to notice a novel road locomotive designed by Mr. Loftus Perkins, and made in 1871 by Messrs. Perkins and Son, Seaford Street, Regent Square London, which was shown at work in the grounds of the International Exhibition at South Kensington in June, 1873. The engine was of the compound type, the diameter of the high pressure cylinder being $1\frac{3}{4}$ in., the diameter of the low pressure cylinder was $3\frac{1}{4}$ in., and both cylinders were $4\frac{1}{2}$ in. stroke. The engine was worked at 450 lbs. steam pressure, and at the time of the Exhibition it ran at a speed of 1,000 revolutions per minute, and had been often at work during the period of $2\frac{1}{2}$ years, and was declared to be in as good condition as when new. The design of the locomotive was somewhat similar to Cugnot's made as far back as 1770 (see page 23 for a description of Cugnot's engine). Mr. Perkin's engine was mounted upon three wheels, a single broad wheel 2 ft. diameter at the front acting as the driving and steering wheel, fitted with a rubber tyre, and two trailing wheels behind. The engine, boiler, and all the machinery was placed on a frame encircling this single driving wheel and turned with this wheel when the stearing gear was actuated. One important feature of the arrangement was, that the engine always pulled in the direction in which it was steered; and all the weight so placed was utilized for tractive purposes. The boiler was constructed of thick wrought-iron tubes with welded ends, the consumption of coal was only 2 lb. per indicated horse power per hour. No exhaust blast was required in the chimney, the engine drew behind it a carriage on which an atmospheric surface condenser was placed, composed of a large number of small tubes into which the exhaust steam was turned. The engine was practically noiseless, and it emitted no smoke, it moved easily at the rate of eight miles an hour, and readily passed over rough places, was steered with facility, and quickly turned about in any direction. This road locomotive was for some time used by the Yorkshire Engine Company, Meadow Hall Works, Sheffield. In October, 1871, the engine drew a wagon load of passengers weighing 33 cwt. from St. Albans to London, 21 miles, at seven miles an hour running time, there being numerous stoppages for vehicles to pass. The indiarubber tyre on the driving wheel was run 1,500 miles without any armour on, and we are informed that no wear was apparent as the engine only weighed 3½ tons. But a special chain armour invented by Mr. Loftus Perkins was prepared and used occasionally. This is the smallest road locomotive we have noticed of the compound type, and this high pressure and high speed miniature engine was said to develop 20 indicated horse power. This was certainly a novel road steamer, but we fear that it was not a practical success.

ARCHER & HALL.—A road steamer invented by Messrs. Archer and Hall was made in 1872 by the Dunston Engine Works Company, Gateshead-on-Tyne, the chief feature of which was the chain armour placed round the indiarubber tyres of the driving wheels. The two cylinders 7 in. diameter, and 10 in. stroke were placed beneath the boiler barrel. The crankshaft, countershaft, and axle were carried on an independent frame of wrought-iron plates extending from end to end of the engine. The feed water was carried in a saddle tank placed above the boiler barrel; and the whole engine was mounted upon springs, so arranged that under all circumstances and conditions, the spur gearing never altered its relation to the several pitch circles.

McLAREN.—Messrs. J. and H. McLaren, of the Midland Engine Works, Leeds, have only been established about 14 years, but in this brief time they have made for themselves a

Fig. 101.

well-earned reputation in the road locomotive business. These eminent engineers have made the latest, and probably the best high speed traction engine ever constructed. In 1885 one engine was built for passenger service in India, which is illustrated by Fig. 101. We quote a few particulars from *The Engineer*: "This engine was of the compound type, and fitted with the well-known spring wheels. It was designed for running at eight miles an hour while hauling a load of 2½ tons. Some difficulty was experienced in finding a piece of road suitable for the test, near the works at Leeds. The shell of the boiler, which was of the ordinary locomotive construction, was made of steel, and the fire-box of Farnley iron, intended for a working pressure of 150 lbs. per square in. The cylinders were 6½ in. and 10 in. diameter, and both of them 12 in. stroke. The crankshaft was 3½ in. diameter, and the main axle 5 in. diameter. Spring spokes were applied to the driving wheels, which were about 6 ft. diameter, and the front of the engine was mounted upon helical springs. There were three travelling speeds provided, the ratios of the steel gearing being 6 to 1, 12 to 1, and 22 to 1. Sufficient water was carried in the tanks for a 20 mile run. The weight of the engine was under 10 tons. A large cab covered the crankshaft, eccentrics, and main bearings; the speed clutch levers were inter-locking, so that it was impossible for a careless engine driver to put two speeds into gear at once. Sectors were fitted to the front axle outside the leading wheels, and the steerage chains were adjusted to prevent back-lash, the steerage action was quicker than usual, a very necessary point for fast travelling; the front wheels had chilled bushes running on case-hardened axle ends. The working parts on the top of the boiler, not covered by the cab, were neatly boxed in to prevent them being covered with dust."

An eye-witness supplied the following account of a run made with the engine previous to its shipment for India. "I wish you had been with us on Monday when the inspector came. We had a grand run; started from the opposite end of Leeds, three miles from the works, about half-past three p.m., and ran over our country roads, where the track was just about the width of the engine, and we had some steep hills—plenty of them one in ten—to go up and down, but no long inclines, some of them were so steep we were obliged to put the slowest gear in to get up. However, the inspector was well pleased with the way the work was done. Just at the finish of the eight miles stretch we had to go down into the valley of the Wharfe, and it is no exaggeration to say that for a short distance the declivity was one in seven, and though we got down very well I could not help wondering what the result would have been if the brake strap had broken. We then had a mile of splendid road, but could not get along at any pace in consequence of hills and traffic; but after leaving the village of Harewood we had four miles of straight road wide enough, but with some long rises of one in twenty. We passed the first milestone in five minutes; the second in seven minutes—having to let a trap pass; the third in six minutes—eased to let carts pass; the fourth in less than five minutes. Running the four miles in twenty-three minutes. The inspector was abundantly satisfied. Three tons was hauled behind the engine, and the tank full of water lasted for the sixteen miles, and would have supplied the engine for four more miles." The omnibus sent to India with the above engine is shown by Fig. 102. We now have pleasure in referring to the three beautiful high-speed road locomotive engines, of excellent proportions, constructed by Messrs. McLaren for the *Fourgon poste* service in the south of France. One of these engines is

McLAREN'S INDIAN OMNIBUS, 1885.

Fig. 102.

shown clearly by Fig. 103.* *The Engineer* for 16th December, 1887, says:—" This service is in the hands of different contractors, and altogether apart from the postal service of the State. It consists of the collection and delivery of parcels and light merchandise in districts remote from railways or indifferently served by them. Strange as it may appear, many of the largest railway centres are also centres of the *Fourgon poste* services, which collect their parcels in one town and convey them by horse conveyance, and deliver them in another town many miles away, although there may be a direct line of railway between the two places. The excessive charges of the railways for goods carried *grande vitesse*, and the excessive time occupied in the conveyance and delivery of goods carried at *petite vitesse* rates, enables these contractors or carting agents to do a large business, many of them requiring several hundred horses for their work."

Some four years ago Messrs. McLaren made one of their compound road locomotives, and tried it for this parcel service with so much success that in a short time others were ordered.

The engines are of the compound type of 12-horse power nominal, constructed for a working pressure of 175 lbs. per square inch. The engine shewn in Fig. 103 is one of the three which are running regularly between Lyons and Grenoble, which are about 70 miles apart. "The goods are collected and packed in the wagon—which will carry about six tons—during the day," and one engine starts out of each town in the evening and delivers its load at the other end the next morning. One engine is kept in the shed in reserve so that the engines can be washed out properly and kept in good running order. The road for about forty miles is very hilly, some of the gradients being

* Kindly supplied by the proprietors of *The Engineer*.

McLAREN'S FRENCH ROAD LOCOMOTIVE

Fig. 103.

1 in 11. There are also some long hills, one of about four miles with an average rise of 1 in 40. "In one part it descends a zigzag course down to the bottom of a very steep valley." Part of the road for several miles runs along the shelving side of mountains, the rocks rising precipitately on one side and the ground falling away on the other side, there is no protection whatever on the lower side, and a moment's forgetfulness on the part of the steersman might plunge the whole train down a precipice 500 feet deep. As the whole journey has to be made in the night "it is of the greatest consequence that the engines should be fitted with ample brake-power and an efficient system of lighting. They are therefore fitted with a steam break—worked by McLaren's patent steam reducing valve—as well as the ordinary hand brake. The former can be applied instantly with such force as to pull the engine up with full steam on, and at the same time, by means of a chain, the brake is also applied to the wheels of the wagon. The engines are fitted with an arrangement for burning ordinary gas. This is compressed into a receiver up to a high pressure, and reduced down to burning pressure by means of a patent regulator or diminishing valve, which Messrs. McLaren specially designed for the purpose. One charging of gas is sufficient to give a brilliant head-light and supply the signal lights for the round trip of 140 miles." The engines are very economical in coal, burning from 5 to 8 cwt. in the trip depending on the condition of the road. The spring arrangement may be described as absolutely perfect, for though the engines have run many thousands of miles there has never been a single breakage in connection with the springs. The roads are not good even where level, as they are full of great holes, and many open drains run across them without any covering. Hugh stones as large as a man's head are constantly rolling down from

the mountains and lodging on the road, so that when the engine comes along at such a speed it has either to drive them out of the way or ride over them, which tries the springs, and is a severe test for shewing the stuff they are made of. A large signal whistle is fixed to each engine. The steersman's seat is mounted upon a spiral spring. The engine work is neatly cased in as shewn, and part of the details are under the cab. Most of the gearing is placed between the wrought iron side plates carrying the crankshaft brackets. The safety valves are spring loaded, the central spring being in compression, out of sight so that it cannot be tampered with. The engines run eight miles an hour for hours together, but of course run slower up hill. Each engine weighs 15 tons fully loaded. The average mileage of each engine is about 15,000 miles per annum. The number on the name-plate of one of these three engines is 288, which shews good work for the time Messrs. McLaren have devoted their energies to road locomotive building. These are splendid examples of high speed compound road locomotives.

And we are pleased that this work contains the description of such a favourable example as the one we have illustrated by Fig. 103; shewing as it does that during the last 23 years (termed the "Modern Period,") considerable progress has really been made in this industry.

The wheel illustrated in Fig. 104 has been recently brought out by Messrs. J. and H. McLaren, and Mr. I. W. Boulton. It is being manufactured by Messrs. John Fowler and Co., Messrs. Aveling and Porter, and Messrs. J. and H. McLaren.

"Referring to the illustration it will be seen that the flange of the wheel is very deep. Square holes are cast in the face, into which are fitted blocks of wood 6½ in. by 9 in., bound with iron, which slide loosely in and out of the holes, just like a piston. These blocks are bedded on indiarubber or felt, to deaden the vibration, and are kept in position by bolts with lock-nuts, which pass through the wheel flange, as shown in the engraving. The blocks are further retained by a stiff spring on each side of the bolts, which has about $\frac{3}{4}$ in.

Fig. 104.

compression. The blocks are set in two rows on the wheel rim, one row half a space in front of the other. The action is as follows:—The blocks in their normal position project a little from the rim, but when the wheel turns, the blocks, coming to the ground, of course yield and present a flat surface to the road. As the wheel continues its revolution, the blocks, rising from the ground, are brought back to their normal positions by the action of the stiff springs, and so the action continues.

It will be seen that the grip of the wood blocks is very great, and no slipping can occur. The fact of the blocks being of wood prevents any damage being done to the road, and the combined action of these blocks and the indiarubber renders the wheels comparatively noiseless.

The wheels now in use, we are informed, have given great satisfaction to their respective owners, the cost of keeping them in repair being much less than ordinary wheels, where the crossbars have to be replaced every few months."*

On paved streets the use of these wheels is specially advantageous, for not only is the tractive power of the engine enormously increased, but any possible damage to the paving setts, caused by the chipping action of ordinary wheels, is entirely avoided. A considerable number of engines furnished with these wheels have been at work in the Manchester district for some years, and the results have been so satisfactory that certain local authorities, who, by virtue of special powers had practically prohibited the use of traction engines within their districts, have waived their restrictions in favour of engines mounted upon these wheels.

Experience has shown, that an enormous amount of wear and tear is occasioned by the shocks sustained by the engines in travelling over rough and uneven roads without springs. The wood-block wheels just described serve as an excellent spring, effectually breaking the shocks, and increasing the durability of the engine. †

* *The Practical Engineer*, January 31st, 1890.
† Steam on Common Roads, by Mr. John McLaren, 1890.

FODEN.—One of the engineering firms which has recently taken up the manufacture of road locomotives is the firm of Messrs. E. Foden, Sons, and Co., of Sandbach. Their first traction engine was turned out in 1880. During this ten years' experience, Messrs. Foden have (in the face of the keenest competition) established a well-deserved reputation, and their road engines have produced some excellent results on the trial field.

The Sandbach firm embodied several wide departures from current practice in their earliest engines, they were not content to run in a groove produced by their predecessors, and the successful records obtained at Stockport in September, 1884, and at Newcastle in July, 1887, proved conclusively that Messrs. Foden's original designs, although at variance with present practice, were based on correct principles, consequently their engines are economical and efficient.

Double cylinder traction engines were often constructed a few years ago, and were generally liked by customers, because they are so easily handled. They do not start so suddenly as the single cylinder engines. They start quietly and run slowly, giving the steersman plenty of time to manipulate the locking, without so much plunging and backing when two or three sharp corners have to be encountered. There is no stopping on the dead centre, occasioning the pulling of the fly-wheel partly round. They are less noisy and rarely prime. But it was generally understood that the double cylinder engine was not so economical in fuel as the single engine of equal power. Messrs. Foden have revived the double cylinder traction engine, thus gaining the above-named advantages, and in spite of the impression respecting its wastefulness, their engines are among the most economical in the market, as was shown at Stockport and at Newcastle.

Messrs. Foden and Co. give the following advantages of the double cylinder:—

First—The strain on the gearing is reduced to a minimum, or less than half required to move the same load with a single engine.

Second—The annoyance of reversing is done away with, which is evidently a great advantage in starting, and is thus more under the control of the driver.

Third—The slow rate at which these engines can be moved gives them an important advantage in hooking on, and being able gradually to start a heavy load, whereas with the single engine the load is started with a jerk, which often breaks the drawbars and pulls the under-carriage-work and numerous other parts to pieces, besides endangering the life of the man hooking on.

Fourth—The strain on the gearing, and therefore on the travelling wheels. being uniform throughout the stroke of the engine, enables it to carry better over soft, greasy ground. The irregular snatching pressure of the single engine we have proved to be one of the prime causes of slipping.

And *Lastly*, but most important—Priming is entirely done away with. by drawing the steam off the water surface gradually ins'ead of by snatches. These advantages equally apply to the compound engines fitted with our patent starting gear.

All Messrs. Foden's traction engines are mounted upon large driving wheels, nearly 7 ft. diameter, the rims are of tough cast iron, shod with steel cross-plates. A peculiarity of these wheels consists in increasing the number of the spokes and using a light section of iron for them. Wheels of such a diameter obtain a great amount of tread, giving them more grip on the surface of the road, thus rendering them much less liable to slip, and allowing the engine to pass over soft ground where wheels of ordinary diameter would sink.

Two travelling speeds are provided, the fast speed ratio of gearing being 16 to 1, the slow speed ratio is 27 to 1.

Fig. 105 shews a traction engine as made by Messrs. Foden, but since the block was cut a few alterations have been made in the design, which must be named. Messrs. Foden are making nearly all their road locomotives with piston valves, they have had them in work two years, and find they answer very well, the wear and tear being practically *nil*, compared with the slide valve. If the feed water in use is dirty, as it very often happens to be, or if the lubrication is neglected, the flat slide valve is a cause of trouble, it wears away at a most rapid rate, and not unfrequently wears the

Fig. 105.

port faces away, which is a serious matter. Hence the number of expedients, which various makers have tried, to prevent this irritating wear and tear.

For instance, Messrs. Aveling and Porter, in 1880, fitted Webb's patent slide valves to some of their engines, this is a round slide valve, which is free to rotate in a hoop, and with improved means of lubrication, some portion of the valve face being always exposed to the exhaust steam. Church's balanced slide valves were used about this time by Messrs. Fowler and others. Everitt's ingenious balanced slide valve, as well as Carter's, have been adopted by some makers. Then the ordinary slide valve has been tried in all kinds of materials, to ascertain the best wearing metal. Brass, phosphor bronze, malleable cast-iron, steel, and hard cast-iron have each been tried, and have their advocates. Phosphor bronze is highly recommended by some traction engine makers.

Messrs. Foden and Sons mount *all* their traction engines on springs. In 1880 they tried flexible wheel spokes, which saved the engines from vibration, but defects soon showed themselves, therefore these flexible spokes were discarded; and in 1882 Messrs. Foden introduced the arrangement as shown by Fig. 106, which answers admirably. It will be seen from the sectional view shewing the patent spring arrangement, that the vibrating shafts are so arranged as not to alter their relative distance, while at the same time allowing the weight of the engine to be carried by the springs. The bearings of the main axle and second countershaft A and B are connected by two levers F, the whole sliding in two axle boxes C, C, preparation being made on the top of the two upper bearings for the reception of the two strong coil springs, contained in the cylinders D, D.

The bearings E E, of the main axle, and the third motion shaft, are of extra length, and parallel, and being coupled by the levers F, having joints at either end, the necessary oscillating or vertical motion is allowed to take place without locking or strain.

The great difficulty hitherto of accommodating the gearing on the stationary to the moving shafts, is overcome in a very simple and effectual manner. The third motion shaft A, which moves up and down, is fixed slightly below a horizontal line drawn through the centre of the second motion shaft, which is not affected by the springs, being fixed in bearings carried by the box bracket, therefore the up

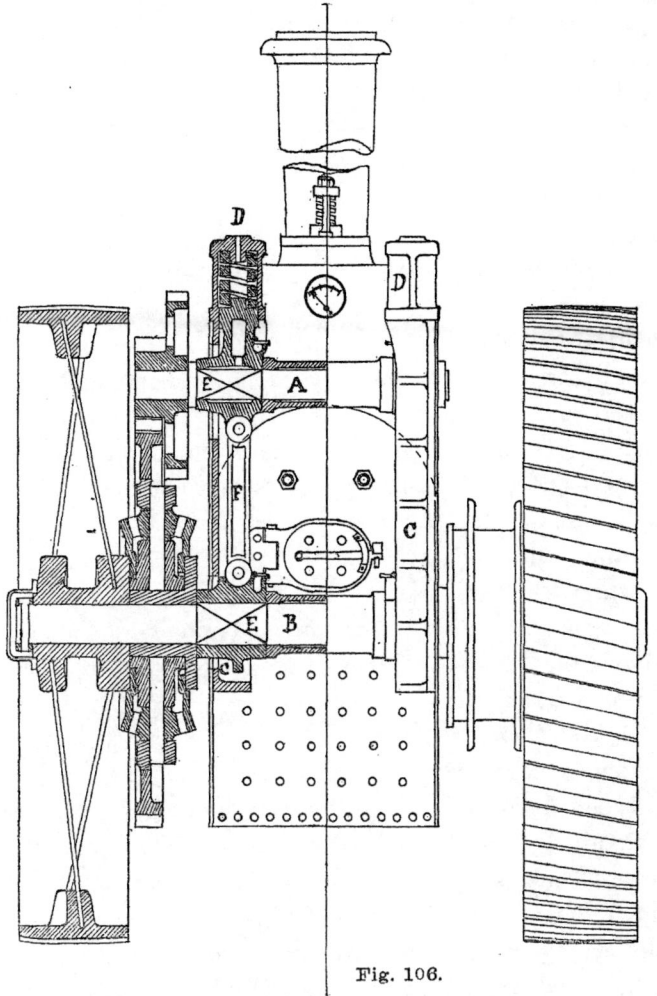

Fig. 106.

and down movement or vibration, which at the most is only half an inch, viz., one quarter of an inch on either side of the centre, does not practically alter the depth in gear of the two spur wheels.

The front part of the engine is provided with a similar spring, on the fore axle, contained in the cylinder, on which are cast stops, to prevent the fore wheels coming in contact with the barrel of the boiler.

Messrs. Foden say: "This perfect spring arrangement materially reduces the effect of shocks or vibrations caused by passing over rough roads, and it is conducive to the reduction of the wear and tear arising from such causes in ordinary traction engines, as leaky fire-boxes, tubes and joints, strained frames and the jolting to pieces of the motion work throughout. Moreover, this spring mounting arrangement adds very considerably to the comfort of the engine driver and steersman."

Fig. 107 represents a sectional view of Messrs. Foden's compound cylinder, showing the steam chests outside. The cylinders and the steam chests are combined in one casting, forming with the jackets a suitable steam dome, containing starting and equilibrium valves, the whole being so arranged as to make priming almost an impossibility. Fig. 107 also shows the new patent compound starting gear, the following being a description:—

The "Compound Engine" is provided with a special arrangement by which the compound action may be instantly suspended, and both cylinders may take high-pressure steam, exhausting directly and independently into the funnel, the steam being supplied in such a manner that each cylinder shall give off the same amount of power. The object of such an arrangement is to give increased power to the engine when starting or doing exceptionally heavy work, as on steep gradients or when getting over soft ground. It is effected in the following manner:—In the passage between the high and low-pressure cylinders, a three-way cock is fitted, this cock being actuated either by an independent lever or else by the starting lever. In ordinary work, the steam from the high-pressure cylinder passes into the larger (or low-pressure) cylinder, is there further expanded, and exhausts therefrom into the funnel.

If, in case of emergency, it is required to get more power out of the engine, the above-mentioned three-way cock is

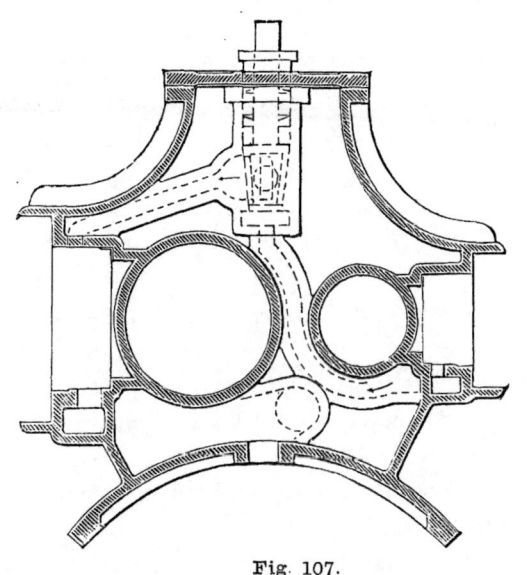

Fig. 107.

opened, so that the exhaust from the high-pressure cylinder passes direct into the chimney, which relieves that cylinder of the back-pressure due to working the low-pressure cylinder, and consequently increases its power. Live steam is at the same time admitted into the low-pressure cylinder; but as this cylinder is so much larger than the high-pressure one, it is obvious that if steam of equal pressure were admitted to both cylinders, the larger one would do the most work, and consequently the engine would run unevenly. To overcome this, a steam reducing valve is provided in the passage to the low-pressure cylinder, by means of which the power to each cylinder is equalised and the engine works as an ordinary double high-pressure engine.

The advantages of this starting gear are : 1st—It enables the user to obtain a great amount of power for starting purposes, getting out of soft places, or taking heavy loads up steep gradients. 2nd—In case of accident to either engine the one may be used independently of the other ; for instance, supposing an eccentric rod broke on either engine all the driver would have to do would be to uncouple both eccentric clips, set the side valve of the disabled engine in the centre of its stroke, and open the three-way cock, as for working double high-pressure ; by so doing the engine can be run as a single high-pressure engine, until such time as it can be repaired. Thus by means of the starting gear, our compound may be converted into double or single high-pressure engines ; the additional complications being only the three-way cock, and the lever for actuating the same.

A very efficient water-heater, forming a suitable foot-plate for oiling and other purposes, is fixed alongside the barrel of boiler, and containing three lengths of steam piping ; a portion of the exhaust steam passing through this warms the feed-water, and what remains uncondensed passes on to the back tank, by which means the greater portion of the heat generated is returned.

The crankshafts are all made of forged mild Sieman's steel turned out of one solid piece, the cranks being of the disc pattern, which greatly assists to balance the motion ; the eccentrics, being also solid with the shaft, and cannot be moved or get out of position in relation to the crank-pins

The six-horse power road locomotive crankshaft is 3¼ in. diameter, and the crank-pins are 3½ in. diameter. Messrs. Foden's recent engines have the feed-pumps fixed to the barrel of the boiler. A circular tool-box for spuds, &c., is fixed on the leading axle.

We must now very briefly refer to the trials of traction engines carried out in September, 1884, at the Stockport Show of the Royal Manchester and Liverpool and North Lancashire Agricultural Society. There were eleven traction engines tested built by seven of the best makers. Each engine was supplied with 20 lb. of fire-wood for lighting up : and 20 lb. of coal per nominal horse power for raising steam, and going through the following manœuvres :—Steam out of the show ground, and traverse a piece of good road sufficiently wide for the engines to turn round, and afterwards enter a clover field cut along the sides for the engines to travel over, then couple each engine to a trolley laden with bales of cotton (the trolley and load weighing about 4½ tons), and run round the field until the fuel was used and the steam exhausted.

It is to be greatly regretted that the question of nominal horse power should have been used by the judges, the inconsistency of which was more than once referred to. One maker called his engine, having an 8½ in. cylinder, a seven-horse power nominal, he was therefore allowed 140 lb. of coal, while two others called their engines eight-horse power nominal with 8½ in. cylinders, and were consequently allowed 160 lb. of coal. During the trials Messrs. Foden's double cylinder locomotive was most carefully driven, and every portion of the heat studiously retained, the blast from the two cylinders caused a more even draught, and the slides had an early cut off, quite independently of the driver minding to keep the reversing lever notched up. This engine consequently made an excellent run. *The Engineer* says :—" Messrs. Foden's engine was more economical than any other engine tried, and they have fairly beaten so far the most eminent firms in the trade. The writer was present at the above trials, and he greatly admired the way the engine performed its task. Messrs. Foden's engine had two cylinders (not compound) 5¾ in. diameter, it was mounted upon large driving wheels, and hung on springs.

The following table gives a few particulars of the two engines Messrs. Foden entered for the Newcastle Trials July, 1887, which were conducted by the Royal Agricultural Society, both of which engines achieved excellent results as we shall show presently :—

TABLE OF
MESSRS. FODEN'S ROAD LOCOMOTIVES AT NEWCASTLE.

PARTICULARS.	SIMPLE ENGINE.	COMPOUND ENGINE.
Diameter of cylinders, in inches	7½	4¾ and 9½
Length of stroke, in inches	10	10
Revolutions per minute declared	168	156
Brake horse power	12	18
Diameter of boiler barrel	2 ft. 6 in.	2 ft. 6 in.
Length of fire-box	1 ft. 9 in.	1 ft. 9 in.
Width of fire-box	2 ft. 0 in.	2 ft. 0 in.
Height of fire-box over grate	2 ft. 2 in.	2 ft. 2 in.
Area of grate, normal, in square feet	3·58	3·58
Area of grate at trial, in square feet	3·00	3·00
Number of tubes	76	76
Diameter of tubes outside, in inches	1·625	1·625
Length of tubes	6 ft. 0 in.	5 ft. 6 in.
Fire-box heating surface in square feet	19	19
Tube heating surface in square feet	177·6	—
Total heating surface in square feet	196·6	201·0
Heating surface per brake horse power	16·75	11·15
Heating surface per square foot of grate	56·15	56·15
Pressure in pounds per square inch	120	250
Coal used per brake horse power	2·555	1·84
Water used per brake horse power	25·63	17·37
Time of running total	—	4 h. 21 m.
Time mechanical hours	4·5	4·583

Messrs. Foden's compound engine worked at 250 lb. boiler pressure. The steam was cut off by a separate expansion slide working on the back of the main slide, the expansion slide gear was a modification of Farcots, and is suitable for the engine running either way. We quote the following from *The Engineer* respecting the workmanship of the engines at Newcastle :—"We believe that Messrs. Foden's works cannot compare in dimensions with those of the great Lincolnshire houses; it is the more creditable to them that the workmanship of the engines and boiler were excellent. Notwithstanding the enormous pressure of 300 lb., the pressure at which the safety valves of the compound engine were set to blow off, the boiler and all its fittings were perfectly tight, not a breath of steam or drop of water being apparent." The engine could be made to work non-compound as per Fig. 107.

We now allow Messrs. Foden to speak for themselves :—
"The results show beyond doubt that by compounding, a saving of at least 25 per cent. is effected. We were competitors in these trials both with simple and compound engines, and in our case the actual saving in fuel by compounding = 29·7 per cent. The benefits of the system do not end here. A corresponding or greater reduction of boiler wear and tear is effected. Again, the consumption of water should not be forgotten; in the above trials the water consumption of our compound engine was only 18·23 lb. per brake H.P. per hour, whilst the best simple engine tried consumed 22·5 lb.

The water-carrying capacity of the tanks is 1,440 lb. exclusive of water in the boiler; this would enable the engine, running at 3½ miles per hour (fast speed), and exerting 18 H.P. actual, to run 4½ hours or 15¾ miles without water; and in confirmation of the above we may say that at the Newcastle trials, our compound engine drove an 18 H.P. load for 4 hours, 22 minutes, on 1,394 lb. of water, whilst the coal consumption was 1·84 lb. per brake H.P. per hour, the highest point of economy of both water and fuel yet attained by any traction engine.

The following table gives a few extracts from the analysis of the waste gases of two of the best engines tried at Newcastle, and published in the *R.A.S.E. Journal*:—

| | SIMPLE. || COMPOUND. ||
	Foden's Traction.	Paxman's Portable.	Foden's Traction.	Paxman's Portable.
Heat units in waste gas (excess of air excluded)	69·924	78·315	87·723	93·910
Heat units in waste gas equal to lbs. of coal	8·42	9·43	10·57	11·31
Heat units lost in sensible heat in excess of air	9·866	96·199	28·799	102·112
Equal to lbs. of coal lost	1·18	11·59	3·47	12·30

The following remarks are quoted from *The Engineer* respecting Messrs. Foden's compound traction engine :— "Concerning the design and the workmanship of this engine we can say that both are as good as that of any other builder of traction engines, and the perfect way in which both boiler and engine dealt with the enormous pressure carried— 250 lb. on the square inch—is sufficient assurance that there is nothing gimcrack about this engine." There are several points respecting the Newcastle engine trials to which we might refer, but we cannot prolong this notice. We have said quite enough to show that Messrs. Foden's engines are worthy of careful consideration.

CONCLUDING REMARKS.

To make a successful steam coach in the early days was no easy task, as we have seen. We believe that had the early promoters spent their time and energies in perfecting an engine *only*, instead of combining the engine and coach in one vehicle, more progress would have been apparent. *The Engineer*, only a few years ago, said: " Anyone can make a steam engine, but only an engineer of special experience can build a successful traction engine."

All the high-speed engines of recent times have been built for service in foreign countries—our foolish and meddlesome laws prohibiting sensible speeds in this country—hence Russia, Greece, Turkey, India, Ceylon, France, New Zealand and Germany are all ahead of Great Britain in this matter. As regards the practicability of running at eight or ten miles an hour on good roads—this is now a settled question. Horses would soon be reconciled to the sight of traction engines on our highways, and if the riders and the childish magistrates could be prevailed upon to cease shying at these engines, some further progress would then be achieved.

ALLCHIN'S TRACTION ENGINE.

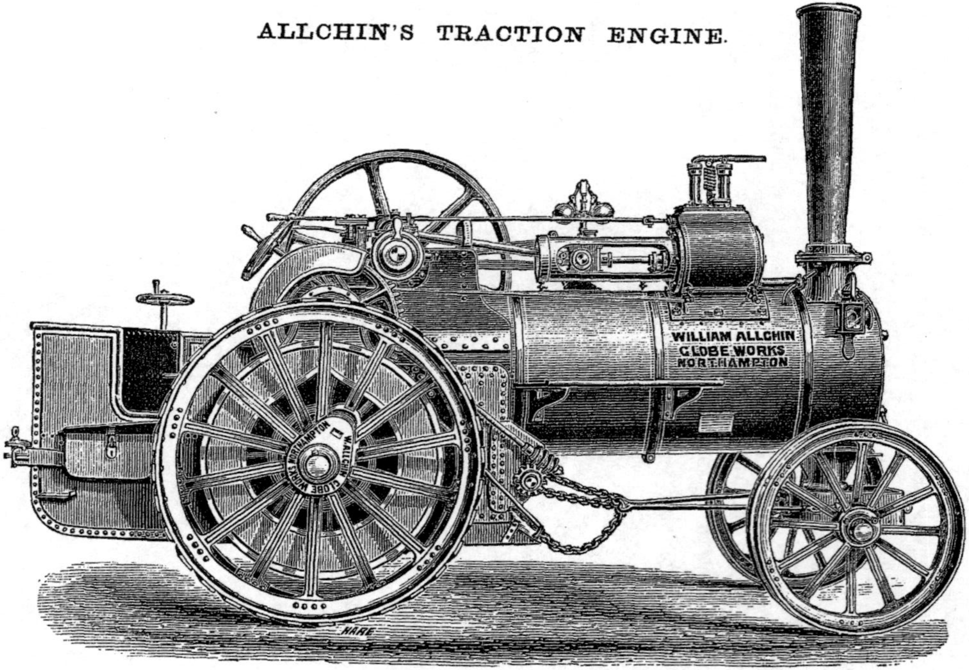

Fig. 108.

PRACTICAL NOTES ON THE DESIGN & CONSTRUCTION OF ROAD LOCOMOTIVES.

Considerable information is already given in this work on the design and construction of road locomotives. The illustrations in the " Modern Period " portray the latest practice of the very best makers, while the dimensions of details embodied therein may be relied upon. Yet, it has occurred to us that the value of the book would be much enhanced by the addition of an equally reliable and complete list of dimensions of single and compound road locomotives.

A table is given on the next page, in which we have embodied the proportions used by the leading makers, supplementing the sizes with others collected by the writer during a lengthened experience. The table gives the data required for designing 6, 8 and 10 horse power nominal both single and compound road locomotives. In addition to this we give some notes which should be serviceable to those entrusted with the " getting out " of this important type of engine.

The general arrangement of all road locomotives is as follows :—The locomotive type of boiler is invariably used. The cylinder is bolted to the barrel of the boiler at the chimney end, while the crankshaft and countershaft carriages are fixed at the fire-box end, the main axle being placed directly underneath the countershaft, below the foot plate, between the fire-box casing and the feed-water tank.

TABLE OF DIMENSIONS OF SIMPLE AND COMPOUND ROAD LOCOMOTIVES.

Nominal horse power	6	8	10
Indicated horse power	30	40	50
Dia. of cylinder, single	8 in.	9 in.	10 in.
Stroke of cylinder, single	12 in.	12 in.	12 in.
Dia. of high-pressure cylinder, compound	5¾ in.	6½ in.	7 in.
Dia. of low-pressure cylinder, compound	9 in.	10 in.	11 in.
Stroke of both cylinders, compound	12 in.	12 in.	12 in.
Working steam pressure, single	125 lb.	125 lb.	125 lb.
Working steam pressure, compound	145 lb.	145 lb.	145 lb.
Revolutions of engine governing, per min.	180	180	180
Revolutions of engine on the road, per min.	240	240	240
Total heating surface, single	120 sq. ft.	160 sq. ft.	200 sq. ft.
Grate area, single	4½ sq. ft.	6 sq. ft.	7½ sq. ft.
Total heating surface, compound	108 sq. ft.	144 sq ft.	180 sq. ft.
Grate area, compound	4¼ sq. ft.	5½ sq. ft.	7 sq. ft.
Diameter of driving wheels	6 ft.	6½ ft.	7 ft.
Width of driving wheels	14 in.	16 in.	18 in.
Slow gear ratio	25 to 1	27 to 1	30 to 1
Fast gear ratio	17 to 1	18 to 1	20 to 1
Diameter of leading wheels	4 ft.	4½ ft.	4¾ ft.
Width of leading wheels	8 in.	9 in.	10 in.
Diameter of main axle	4½ in.	5 in.	5½ in.
Diameter of crankshaft	3¼ in.	3½ in.	3¾ in.
Diameter of crankpin	3½ in.	3¾ in.	4 in.
Pitch of first motion spur gearing	1¾ in.	2 in.	2¼ in.
Pitch of second motion spur gearing	2 in.	2¼ in.	2½ in.
Pitch of third motion spur gearing	2¼ in.	2½ in.	2¾ in.
Diam. of compensating gear bevil wheels	2¼ ft.	2½ ft.	2¾ ft.
Pitch of compensating gear bevil wheels	2¼ in.	2½ in.	2¾ in.
Slow travelling speed in miles per hour	2½	2½	2½
Fast travelling speed in miles per hour	4 to 5	4 to 5	4 to 5
Diameter of Fly wheel	4 ft.	4½ ft.	5 ft.
Width of Fly wheel	6 in.	6 in.	6¼ in.
Weight of engine in working trim	10 tons.	12 tons.	15 tons.
Approximate load hauled on fairly level roads	18 tons.	25 tons.	32 tons.
Total width of engine	6 ft.	6¼ ft.	6¾ ft.
Total length of engine	16 ft.	18 ft.	18½ ft.

The cylinder should be thoroughly steam jacketed, and well drained; it should be bolted on the boiler barrel as near to the smoke-box as possible, and be provided with a flange all round the base, the bolts in which should be about 4 in. pitch, sufficient bearing surface being allowed to make a joint under the foot; the steam inlet is usually placed close to the smoke-box tube plate. Ample room must be provided round the liners, and in the stop valve chamber, to form a reservoir or dome, so as to prevent priming. It is important that the steam passages and ports should be of sufficient area to admit of a high piston speed, and allow the steam to follow the piston at the necessary velocity. "For this reason small ports are useless, as when the link is notched up and the travel of the slide valve thereby reduced, the openings are too cramped for the steam to pass in and out of the cylinder freely; the result is that the slide valve is forced off the face and the engine primes as soon as any great speed is attained. It is easy to tell by the sound of the exhaust if the passages are rightly proportioned, and whether running at high or low speeds the engine should give a clear, distinct beat." The stop valve and equilibrium throttle valve may be contained in the upper part of the cylinder casing, and the Ramsbottom safety valves are conveniently placed on the top cover.

It is usual to allow 10 circular inches of piston area for each nominal horse power in the single cylinder engine, hence the sizes given in the table on the previous page.

"The practice of making road locomotives upon the compound principle is increasing; where coal is expensive, or water scarce, compound engines have a decided advantage, as in favourable circumstances an economy in fuel of as much as 30 per cent. may be effected. Moreover, the shocks upon the crank pins, gearing, and other working parts are less severe, and as the steam is expanded down nearly to atmospheric pressure, the noise of the exhaust is much reduced, and the danger of frightening horses considerably diminished." The best ratio of cylinder capacity has been found in practice to be about 2.4 to 1.* Therefore we have adopted this proportion in deciding the diameters of the compound engine cylinders in the table; and it will be noticed that these sizes agree with dimensions given in previous pages of the book. The steam chests should be placed outside, so as to be easy of access for examination or repairs of the slide valves. An efficient auxiliary valve should be provided, which allows high pressure steam to enter both steam chests, and both the cylinders exhaust into the common exhaust pipe, the compound principle being suspended for the moment, for starting with a heavy load, ascending a steep hill, &c. A relief valve must be fixed on the low pressure steam chest.

For such high working steam pressures the boilers are usually made of steel. The fire-boxes in some cases are made of Lowmoor or Bowling plates. All the circular seams are single riveted, while the longitudinal seams are double riveted, the fire-box sides and roof stays are pitched 4 in. centre to centre. Solid iron fire-box foundation and fire-door rings are used, instead of Z iron. The water space round the fire-box is made 3 in. at the bottom. Four mudholes must be provided in the most convenient positions. If the under carriage for the front axle is attached to the smoke-box, the plate should be $\frac{3}{8}$ in. thick in this part. The barrel plates are $\frac{3}{8}$ in. thick. The fire-box sides carried up to form the brackets are $\frac{3}{8}$ in. thick. The front, back, and arch plate of the fire-box shell are $\frac{7}{16}$ in. thick. The inside fire-box plates are $\frac{3}{8}$ in. thick. The fire-box tube plate is $\frac{5}{8}$ in. thick, while the smoke-box tube plate is $\frac{9}{16}$ in. thick. The boiler is machine

* "Steam on Common Roads." Mr. John McLaren, Assoc. M.I.C.E.

riveted all over, and must be equal to the very best railway locomotive work, for in addition to having to bear a working pressure of 145 lbs. per square in., a margin of strength must be allowed to resist the additional strains from the machinery. For boilers intended for single cylinders we have allowed 20 square feet of heating surface per nominal horse-power, and ·75 square feet of grate area per nominal horse-power. For the compound engine boilers 18 square ft. of heating surface is allowed per nominal horse-power, and ·70 square ft. of grate area. The boiler should be provided with two gauge glasses, a sliding fire-door, and a smoke-box door that can be tightened equally all round.

The box brackets should be formed of the fire-box sides carried upwards, and two cross plates for stiffening purposes; these cross plates are riveted to the arch-plate by means of angle iron, in addition to these two transverse plates some other means of stiffening the side plates will be required; for instance, Messrs. Burrell use a countershaft carriage extending across. Messrs. Aveling employ a fixed intermediate shaft, others use a plate at the top of the side plates, forming a tray over the shafts, &c. The carriages for the crankshaft, countershaft and main axle must be let into the side horn plates, the bolts or rivets merely keeping them in place. These carriages are often made of cast steel so as to reduce the weight. The box brackets must be machine riveted all over.

"The tender and hind tank should be bolted to the horn plates by turned bolts, independent of those in the main axle brackets, so that when necessary the tender can be removed, without disturbing any other part." The sides of the tender are stiffened where the attachment is made to the horn plates. " The drawbar should be arranged so as to transmit the pull of the engine direct to the horn plates, and avoid any strain upon the plates of the tank, tending to make it leak." *

A spring draught bar for coupling the engine and the wagons has been used, and proved itself to be of great service in saving wear and tear, caused by suddenly starting and stopping, and due to the irregularities of the road also.

The driving wheels are usually made by putting two wrought-iron rings of tee-iron side by side (welded and blocked to a true circle on a level plate), with steel cross plates riveted directly to the tee-irons as shewn by Fig. 93. The tee-irons for wheels 16 in. on face are 8 in. wide, 4 in. deep, 1 in. thick on the web, and $\frac{7}{8}$ in. thick on the outside. The spokes are cast into the boss, and have tee-ends welded to them at the outer end, which are firmly riveted to the web of the tee-iron ring by hydraulic machinery.

Another method, shewn by Fig. 106, is sometimes adopted, the spokes are cast into the boss at one end, and into the cast-iron tyre at the outer end; the cross plates are then riveted to the tyre. The brake barrel is occasionally cast to one of the wheel bosses, so that it cannot be rendered useless by the driving pins coming out.

"The best arrangement for the fore carriage consists in mounting the fore axle beneath a cast-iron bearing, which forms a sort of turntable, supporting the front end of the boiler. The axle is carried beneath by a horizontal pin, upon which it rocks, so that there is a sort of universal joint under the front end of the engine, permitting the axle both to tilt vertically, so that the wheels can accommodate themselves to any inequalities of the road, and also turn in a horizontal path, to such an extent as may be required for

* "Hints to Purchasers of Road Locomotives," by Messrs. Charles Burrell and Sons, 1890.

the purpose of steering the engine. The steering gear is generally worked from the foot plate by a worm and wheel at the end of the transverse horizontal shaft, around which are coiled double chains—one winding and the other unwinding as the shaft revolves. The front axle is thus drawn into and held in whatever position is required for turning the engine, or for travelling straight forward."*

The steering gear hand wheel is usually placed on the fly-wheel side of the foot plate. A box for the spuds is fixed on the front axle.

Every road locomotive is now fitted with compensating gear on the main driving axle, as clearly shewn by Fig. 93. This gear consists of two bevil wheels and two bevil pinions arranged as follows:—The right hand side bevil wheel is bolted or riveted to the driving wheel boss, which runs loosely on the axle. Another bevil wheel on the left hand side is keyed to the axle. The two pinions are carried on studs fixed in the compensating plate. The spur ring of the road gear is riveted to this plate; when the engine travels in a straight line the teeth of the pinions act as drivers—the pinions do not revolve on their studs—but drive both the bevil wheels at the same speed; when the engine is required to turn to the right or the left, one driving wheel having a tendency to travel faster than the other, the bevil pinions revolve on their pins to allow for this accelerated speed of one wheel. But Fig. 93 shews a method of locking the compensating gear when required; a pin is placed in the hole shown in the wheel boss, which projects into the boss of the centre plate; it will be seen that when the pin is inserted the bevil wheels can only then act as drivers. This locking gear is only required occasionally, for instance, when

* Steam on Common Roads by Mr. John McLaren, Assoc. M. Inst. C. E.

one wheel of the engine gets on to soft ground, or when the engine is mounting a hill with a heavy load behind, when the road is slippery, one wheel will do all the work the other merely spinning round and slipping. The writer was lately accompanying an 8 horse power road locomotive which was hauling a load of 20 tons up a hill having a gradient of 1 in 13 and in one place of 1 in 11, the road was hard and slippery, one wheel was constantly slipping on an icy surface, but after locking the compensating gear the engine ceased to slip and proceeded at a good rate to the top of the hill. We give the sizes and the pitch of the compensating bevil wheels in the table, on page 260.

On the fly-wheel side of the engine a winding drum is keyed to the axle (see Fig. 93), this drum drives the left-hand side wheel, by means of a strong driving pin. "By withdrawing this pin the wheels may be thrown out of gear, and the winding drum made to revolve upon the main axle, without moving the engine; the whole power of the engine may thus be concentrated upon this drum, round which a steel wire rope can be used with great advantage. For instance, the engine may be loaded with a heavy casting or other weight, up to the limit of its power with the slow gear. It may be able to haul this load along all ordinary roads where there are no heavy gradients; but quite unable to mount with it up a steep hill. In the latter case the engine would be detached and moved up the hill, the load being left at the bottom; the road wheels would then be thrown out of gear and firmly 'scotched,' and the wire rope would be made fast to the load, which could then be wound to the top of the hill with the greatest ease. It is evident that this arrangement of drum and rope may be made to serve a variety of useful purposes."[*] Messrs. Burrell and Sons, some years

[*] "Steam on Common Roads," by Mr. John McLaren, A.M.Inst.C.E.

ago patented an arrangement of winding drum combined with the compensating gear, one advantage of which is, that the main axle does not revolve when the winding drum is in use. This combined drum and differential gear is used by several firms. The steel wire rope can be paid out as the engine travels forward, if the guide rollers are fixed in a line with the underside of the winding drum.

The steel gearing is so arranged that two travelling speeds are provided. Pinions of two different diameters are fitted on the crankshaft, which gear into two spur wheels suitably proportioned on the countershaft. The countershaft carries a pinion which gears into a wheel on the axle. We have given the ratios of the gearing in the table—that is, the number of revolutions the crankshaft makes to one of the driving axle, in the fast and slow gear.

This ratio is a very important subject, but we have dealt with it under Messrs. Hornsby's name on page 233, so we cannot re-open the question here, but we may say that we have arranged the gearing ratios so that the engine travels 30 per cent. faster in the quick speed than the slow. In bygone days there was generally 50 per cent. difference in the speed, rendering the fast gear comparatively useless. The pitch of the gearing is given in the table on page 260.

This brings us to the subject of the number of shafts, and the number of wheels used to fill up the interval between the crankshaft and the spur wheel on the axle. The engines are known as three-shaft engines, and four-shaft engines. Messrs. Aveling and Porter's road locomotives are four-shaft engines; the gearing is shown by Fig. 92, page 218.

Messrs. Chas. Burrell and Sons use three shafts and they refer to the subject as follows :—" Simplicity of construction is a most essential feature in a traction engine, more especially as regards the arrangement of the gearing, and the fewer the number of shafts and cog-wheels the better, in order to save wear and tear and reduce friction and superfluous weight." Fig. 84 shews a three-shaft engine. Fig. 100 shews the first motion gearing of a three-shaft engine, which has some valuable advantages as named under Messrs. Hornsby's notice. The four-shaft engines possess some advantages which we have noted, and that cannot be obtained in any other manner, but we fear that these advantages may be too dearly bought, as the friction is undoubtedly increased by the use of an extra shaft, with its bearings, its gearing and the increased weight. The sides of the box brackets have likewise to be made longer in order to receive the extra countershaft bearing.

When experts differ, who then shall decide? We know that Messrs. Aveling have good reasons for their use of four shafts. And Messrs. Burrell have ample grounds for being satisfied with their choice of the three-shaft arrangement. Time alone may ultimately settle the question.

The crankshafts are usually bent out of a round bar of mild steel; bent cranks with the dips made square are sometimes used. The collars should be forged out of the solid material. It is usual to make the crank pins larger in diameter than the crankshaft. The keys for the driving pinions should be cut out of the solid material (for we know how soon the sunk keys or feathers work loose), all risk arising from badly fitted keys is thus avoided.

We may here refer to Messrs. Foden's crankshaft, the cranks of which are discs; the eccentrics are turned out of the solid material.

The countershaft, or countershafts, as the case may be, should be of large diameter and made of steel.

The bearings for all the shafts should be in length at least twice the diameter of the shaft. The carriages for the shafts are of wrought iron or steel, and they are riveted to the horn plates. Phosphor bronze bearings are often used with good results.

"The connecting rod should have the strap securing the brasses bolted through by two bolts, and have one cottar to adjust the brasses. This is a more solid job than when the old-fashioned gib is used," and to prevent the cottar from coming out, a grove about one-sixteenth of an inch deep should be provided on the side of the cottar, into which the end of the set screw works. The connecting rod should be $3\frac{1}{2}$ times the length of the stroke of the cylinder, so as to reduce the angular pressure upon the slide bars.

The link motion reversing gear must be most carefully designed. The slip upon the die should be reduced to the smallest possible amount, and the parts ought to be so arranged as to give an equal cut-off and release at both ends of the cylinder, whether working in full gear or notched up. The engine should exert the same power whether working in forward or backward gear, and the cut-off should be as sharp as possible.

All the pins and wearing parts must have plenty of surface and be deeply case-hardened. The reversing gear lever is usually fixed at the right hand side of the foot-plate. The weighbar shaft and valve rod guide should be so arranged that long eccentric rods can be employed, for reasons we have already stated. Let the weighbar shaft bearings be adjustable for taking up the wear.

The lubrication of the slide valve is a very important matter; the oil should be conducted on to the valve face, and not allowed to run down the steam chest side and miss the parts requiring lubrication. We have spoken of slide valves under Messrs. Foden's name.

Cast iron eccentric straps are always used on the best constructed engines. We quote the following from *Engineering*: "Cast iron eccentric straps, when properly got up, and made a slack fit top and bottom, with all the corners carefully rounded off to a large radius will, with proper oiling, run almost without wear. Brass straps wear whether oiled or not, and if neglected simply get warm and wear faster.

"Given a careful engine driver, cast iron is therefore far the preferable material. Customers, however, are prejudiced against them, because they have an idea that the engine makers use them because they are cheaper than brass ones; but this is not the reason why they are adopted. Cast iron is a better wearing material than brass. The eccentric rods have tee ends for bolting to the eccentric clips. Flat eccentric rods and link motion details look much nicer than round ones, but the former cost a trifle more than the latter."

The feed pump is often placed on the boiler barrel as shewn in Fig. 95. They were put in this position when the inside gearing was introduced. The main disadvantage is that they are out of the reach of the engine driver when travelling on the road, should anything go wrong. But it is well known that pumps placed on the boiler with a short inlet pipe are less noisy than those fixed on the box brackets or on the countershaft bearing, and having to force the feed water through a long pipe between the pump and the check valve. When they are fixed on the boiler barrel, a wrought iron or cast steel seat should be riveted to the boiler to receive them, the seat and the pump faces being planed, so that the pump can be readily taken off without breaking an awkward joint. The check valve should have a clearing-out hole opposite the inlet pipe. It is usual to fit the pump with a solid plunger. the joint being outside and easily oiled. Continuous pumps are generally adopted, so that the water is always circulating. Some makers use ball valves for the pump and the clack valve; others use spiral wing valves. It is not a difficult matter, however, to so arrange the pump boxes that either kind of valves may be used if desired. All joints in the suction and other pipes should be made with round flanges, so as to be easily taken down for examination. The water in the tank under the boiler barrel is usually heated by the hot condensed steam from the exhaust chamber of the cylinder. In addition to the feed pump a non-lifting injector is employed on road locomotives, so that water may be forced into the boiler when the engine is standing on the road. The injector is usually fixed near the bottom of the hind tank so that the water flows into it, the cocks being arranged handily for the engine driver. A water-lifter and 25 yards of 1½ in. hose pipe is always fitted to road locomotives, so that the tanks can be readily filled from a stream on the road-side. The water lifter is easily placed on the top of the fore tank as shewn in Fig. 97. The hind and fore tanks are connected by a pipe so that the water in both of them maintains the same level, a cock being inserted in the pipe.

It is very important to have as few holes drilled in the boiler as possible, hence the one bracket used for carrying the governors is made to support the slide bars, in some cases, also, the valve rod guide, the weighbar spindle, the governor driving gearing, the throttle valve spindle, and occasionally it is made to support the stop valve rod. In order to further reduce the number of 'bits,' and lessen the quantity of the holes in the boiler, the water-filling hole is arranged in the cylinder foot, see Fig. 87. The steam blower pipe can be taken from the jacket of the cylinder, instead of screwing a bent piece of pipe into the boiler. The whistle, pressure gauge, and the injector receive their steam from the same fixing on the boiler.

The governors should be of the spring type; they are, of course, only put on road locomotives in case the engine is intended to drive machinery from the fly-wheel, when not at work on the road. *In no case is the governor ever used when the engine is travelling.* Need we specify that the attachment from the governor to the throttle valve should be simple and direct, not two pins to be used when a little scheming would make one pin answer the same purpose. Some of the old traction engines were made to rejoice in a superabundance of joints, bell-cranks, pins, and links for connecting the governor with the throttle valve. Suppose the opening of the equilibrium throttle valve has a lift of ½ in., the back lash and the inevitable slackness of two or three pins and joints partly worn, soon renders the governors comparatively useless. Hence the reason for direct acting gear. In the table we give the speed the engine will run when the governors are in use, and the approximate speed when the engine is running on the road, and the governors are not at work. See table on page 260.

The stop valve lever is attached to the front plate of the box brackets, the rod running parallel to the centre line of the boiler barrel, high above all the details of the engine. The valve is fixed in the highest part of the cylinder, the valve has a V cut in the face, so as to admit the steam gradually for starting. A displacement lubricator should be fixed over the stop valve to lubricate the stop and throttle valves; another should be provided for oiling the piston. These lubricators should be of the displacement type, having a plug in the bottom, so that they can be shut off for filling when under steam, and by turning the plug in another direction the oil can be fed into the cylinder as fast as required. For compound cylinders one lubricator is used above the throttle valve, another is fixed over the high pressure slide valve, and one more feeds oil over the low pressure slide valve. Englebert's lubricant is one of the best cylinder oils. If once used will always be preferred to any other.

In lieu of two or four slide bars, bored guides are often used as shown on Fig. 105. Messrs. Allchin's engine, Fig. 108, on page 258, has bored guides. Two angle iron slide bars are adopted on Fig. 97. The crosshead in this case is forged on the piston rod.

Spring balance safety valves are not generally used, but spring weighted safety valves are mostly adopted, having one central spring in tension (Ramsbottom type). Another maker uses a central spring in compression, while several other firms use a spring in compresssion over each valve, all these types are certain in action, and cannot be tampered with by the engine driver, as the spring balance type may be.

All the handles for starting, stopping, and reversing the engine, for opening and closing the cylinder cocks and the auxiliary valve, for starting the pump or the injector, for opening the sliding firedoor, for putting the road gear pinions in and out of gear, for opening and closing the ashpan damper doors, for putting on the brake, for testing the height of the water in the boiler, for sounding the whistle and for

T

steering should be placed as handily as possible, so that one man may drive and steer the engine if required for short intervals.

A throttle valve damper in the chimney is a useful detail but not often adopted. The joint on the chimney base should be placed on the front side, so that the chimney can be lowered down across the smoke-box door, for passing under low bridges or into sheds.

Steel and wrought iron should be used freely in order to gain strength with lightness, bolts in the place of studs should be employed, rivets can be used to advantage in many positions, locknuts must not be overlooked where the bolts are subjected to much jolting

" A sufficient amount of weight on the driving wheels is required to absorb the full tractive power of the engine, but excessive weight upon the front wheels increases the friction of the steering gear, and causes the front wheels to "burrow" amongst loose material on the road. It is therefore advisable to put the bulk of the weight upon the hind wheels (which may be made wide enough to carry it), leaving sufficient weight only upon the front wheels to enable them to control its direction." *

The superiority of steam over animal haulage is now universally admitted. A saving of from 40 to 60 per cent. can be affected by the employment of the best classes of steam machinery.

The cost for haulage may be put down at 1d. to 3d. per ton per mile.

* Steam on Common Roads by Mr. John McLaren, Assoc. M. Inst. C.E.

Mr. John McLaren gives the following table showing approximately the gross and nett effect which can be obtained from their engines, on macadamized roads in fair condition. The data in this table refers to single cylinder engines, cutting off at about ¾ stroke.

Nominal Horse Power.	Weight of Engine (empty), in tons.	Weight of Engine in working order, in tons.	Steam Pressure, in lbs., per square inch.	Indicated Horse Power at 400 feet, piston speed.	Number of Wagons in train.	Weight of empty Wagons, in tons.	Net load (contents of Wagons), suitable for level roads.	Gross load (Engine, Wagons, and contents), on level.	Net load on incline of 1 in 10.	Net load on incline of 1 in 10 (extra good roads.)
6	9	10	130	25	2	5½	12	27½	9	10
8	10½	12	130	35	2	5½	16	33½	12	14
10	13	15	130	55	3	8¼	21	44¼	18	21

THE FOLLOWING TABLE OF DIMENSIONS OF TRACTION ENGINES HAS BEEN PUBLISHED BY A WELL-KNOWN LINCOLN FIRM. IT MAY BE USEFUL.

	SINGLE CYLINDER.				DOUBLE CYLINDER.	
Nominal H.P.	6	7	8	10	8	10
Dia. of Cylinder	8 in.	8½ in.	9 in.	10 in.	6¾ in.	7½ in.
Stroke	12 in.	12 in.	12 in.	12 in.	12 in.	12 in.
Dia. of Flywheel	4 ft.	4 ft.	4 ft 3 in	4 ft 3 in	4 ft 3 in	4 ft 3 in
Face	6 in.	6 in.	6 in.	6 in.	6 in.	6 in.
Revs of Crankshaft per min.	150	150	150	150	150	150
Heating Surface, square feet	20	131	157	182	157	182
Water required per hour, gals.	75	90	100	125	100	125
Good Coal used per hour, lbs.	96	112	128	160	128	160
Water carried, in gals.	115	126	138	174	138	174
Coal carried, in cwts.	5½	6	7	8	7	8
Hours run without coaling	6	6	6	6	6	6
Do. do. watering	1¼	1¼	1¼	1¼	1¼	1¼
Total full weight, in tons	9½	10	11¼	12½	11¾	13
Total empty weight, in tons	8½	91	10	11	10½	11½
Speed, slow, miles per hour	1½	1½	1½	1½	1½	1½
Do. fast do.	2¾	2¾	2¾	2¾	2¾	2¾
Weight hauled on good road at fast speed, in tons	15	17	20	25	20	25
Load up incline, 1 in 12, tons	7	8½	10	13	10	13
Weight on each hind wheel, in tons	3	3½	3¾	4	3¾	4
Dia. of Driving wheel	5 ft 6 in	5 ft 9 in	5 ft 9 in	6 ft.	5 ft 9 in	6 ft.
Face of do.	14 in.	16 in.	16 in.	18 in.	16 in.	18 in
Dia. of Leading Wheel	3 ft 4 in	3 ft 6 in	3 ft 6 in	3 ft 6 in	3 ft 6 in	3 ft 6 in
Face of do.	9 in.	9 in.	9 in.	10 in.	9 in.	10 in.
External Length	14ft 8in	15 ft.	15ft 6in	17ft 4in	15ft 6in	17ft 4in
Width over Wheels	6ft. 9in	7 ft 1 in	7ft 3½in	7ft 7½in	7ft 3½in	7ft 7½in
Circle to turn in, in feet	25	26	27	29	27	29
Approx. Weight packed, cwts.	190	205	225	240	235	250
Approx. Cubic feet	500	580	730	860	730	860

The weights hauled and coal consumed will vary notably according to the state of the roads and the skill of the driver.

When travelling on the road, drivers often throw off the governor belt and run the engine faster, the speeds will then of course be greater. When taking a light load the coal and water consumption will not much exceed one half the quantities named above.

BEAUTY OF DESIGN.

We trust that the notes given on the preceding pages of the book may serve as useful hints on the design of road locomotives. It is impossible to do more than touch upon the fringe of this great subject; a chapter could be written on each topic. By way of conclusion we may offer a few remarks on THE BEAUTIFUL IN MACHINERY, having reference to the type of engines under discussion. Our best engineers, when designing, aim at securing a symmetrical appearance as one of the advantageous features of their productions, and the time spent is not lost, but is bearing fruit. Because ribbed and angular details of road engines are bad and ugly, they are gradually giving place to hollow castings, which are far more graceful in external appearance, and equally strong, besides possessing two other good features viz., being easily painted and kept clean. Some of our modern engines are very pleasing to the eye, they are most carefully balanced in the arrangement of their details, and all curves are drawn as gracefully as possible. But unfortunately all persons are not able to appreciate beauty of form in machinery when they see it, and do not aim at its attainment. If they draw a fairly neat looking detail it is more the result of accident than design.

Beauty of design is more easily appreciated than described. It consists of no alteration in the principle, neither does it affect the internal details of the engine, but it is brought about by an expenditure of drawing-office care in the arrangement of the parts, which gives to the whole a symmetrical and *simple appearance*.

Every detail is made to possess perfect and graceful proportions and a pleasing outline, and the shape of one part is not allowed to be out of accord with any other part, and in no case is *real efficiency* sacrificed thereby.

It is matter for congratulation that some of the modern road locomotives shew a marked improvement in their design. While the 'cardinal virtue' simplicity, or fewness of parts, is being constantly aimed at with the best results. It is not many years ago that some of the traction engines we saw on the road, looked like a part of an engineer's factory out for an airing. A few years ago *The Engineer* spoke of traction engines as being the quintessence of all that is ugly, noisy, and in very deed a nuisance. The same authority went on to remark :—" We have an engine without springs, and with all the gearing exposed to the broad glare of day, thumping, and clanking, and grinding, and smoking along our highways. The thing is a nuisance, an unmitigated nuisance, and it is folly to deny the fact."

The above remarks in no way apply to the road locomotives made by the ablest makers of the day, which are fully described in this book.

In getting out a design a good appearance should be studied ; it is just as easy to produce a well proportioned engine as an ugly one.

Ugliness and awkwardness are generally companions, while the best looking engines are invariably the best ones to go. The design is an index to the character of the whole engine. If the designs are good, no misgivings need be entertained respecting the quality of the materials employed, or the workmanship.

An appreciation of, and a strong desire to produce graceful forms and dainty curves is being fostered in some drawing offices with success. Finality in design has not been attained yet. Our road locomotives shall yet possess a good name for quietness, efficiency, and beauty of appearance.

ROAD LOCOMOTIVE

OR

TRACTION ENGINE LAW.

We purpose drawing the reader's attention to several unreasonable and almost prohibitive clauses in the Road Locomotive Acts of 1861, 1865, and 1878, now in force.

"Among the legal restrictions may be mentioned, firstly, that such engines have been held to be a nuisance at common law; secondly, that in nearly every case the owner is obliged to obtain a licenee from a Court of Quarter Sessions, before he can travel with his engine on any highway, and that he may have to wait nearly three months before such licence can be granted to him; that this licence, though it is evidence that his engine is constructed in accordance with the requirements of the Act of Parliament, affords the owner no protection whatever against any person or public body raising the most frivolous objections to the passage of his engine. Thirdly, that in country districts, though there may be no other traffic on the road, the speed of the engine is limited to four miles an hour, and a man is required to walk in front at a distance of not less than 20 yards. Fourthly, that the road authorities have an almost arbitrary power to forbid the use of certain bridges by such engines, though the bridges themselves may be of ample strength to carry the weight without danger; and further, that although the damage done to a bridge by the passage of heavy weights, drawn by horses, is made good at the public expense, such damage must be made good by the engine owner, in case the same load should happen to have been drawn by a traction engine. Fifthly, that certain urban authorities have been allowed to embody in their local Acts clauses by which they are able to prohibit the use of road locomotives on any street or road within their jurisdiction."* There are other restrictions which we shall notice in due course.

There is something unaccountable in the prejudice against the use of steam on common roads which obtains in this country, and also in the success which has attended the exertions of those, in whose sight any power save that of horses is an abomination, who propose to abate what they are pleased to style a nuisance by the aid of legislative enactments.

Ever since the introduction of steam locomotion on common roads, there has existed a considerable number of influential persons who have waged a terrible crusade against these engines traversing the roads of their districts. These injurious Acts of Parliament, have given local busybodies and biased county authorities the power to place all sorts of obstructions in the path of one of our most important agricultural and industrial pursuits ; and, we need scarcely add, that this power has been freely and repeatedly used, as the owners of traction engines all over the country know too well. The influential party who are inimical to the use of locomotive steam engines have become so active and merciless that to move an engine from one part of a district to another is to run the risk of a summons for some possible or inevitable infringement of the Acts, as interpreted by these opponents

* Steam on Common Roads by Mr. John McLaren, Assoc. M. Inst. C. E.

to their use.* This is veritable persecution, to which all owners of such engines are liable so long as the above-named Acts are in force, for under these laws it is impossible to construct locomotives fit for the work required of them without infringing one or other of their clauses.

Considering the immense amount of money invested in the thousands of self-propelling engines used, and the agricultural machinery employed therewith, it is somewhat surprising that a united effort has not long ago been organized by the owners and manufacturers of these useful engines, for the purpose of removing these oppressive Acts, which are such a serious obstruction to our agricultural and commercial interests.

The first unreasonable regulation we shall notice is the one pertaining to steam blowing-off on the road, which runs as follows :—" Nor shall the steam be allowed to attain a pressure such as to exceed the limit fixed by the safety valve, so that no steam shall blow off when the locomotive is upon the road."

It is impossible for traction engine drivers to comply with this law so long as the engines are compelled to stop for conveyances to pass. Here is a case in point :—the driver has just replenished his fire in readiness for mounting a hill or traversing a heavy piece of road ahead, and as the fire brightens up and the steam approaches the maximum pressure,

* " Laws were actually passed in England, on the first introduction of steam on railways, limiting the pressure in the boilers to 30lbs per square inch. The first railroad charter contained a clause limiting the speed of trains to 12 miles an hour, and when 30 miles an hour was suggested, it was ridiculed as an idea simply insane. And it was said people would just as soon be persuaded to allow themselves to be fired out of a cannon as to be hurled along at such fearful velocities, which would, without doubt, have the most disastrous effects upon the circulation of the blood, and other vital organs."—*Industries*, 5th September, 1890.

And so long as we place power in the hands of people who have no practical knowledge and very little wisdom, that power will be used to stifle progress. The same class of individuals who opposed railways, now oppose road locomotives.

at that precise moment, unfortunately, a carriage and pair of horses appear in the distance. The engine is stopped to allow the equipage to pass, which, however, with the horses generally harnessed to gentlemen's carriages, is no easy matter, and while these foolish animals are prancing and plunging, instead of passing along the road, the pressure gauge pointer on the engine travels fast, notwithstanding that the driver has closed the damper door, and opened the fire door ; and the only alternative is to allow the steam to blow-off and thus break the law, or keep the law by wedging down the spring balance to a dangerous degree, provided the engine is fitted with spring balances, which we are pleased to state has lately become exceptional. Most of the principal makers now fit their engines with two spring-loaded safety valves, which cannot be tampered with by the driver ; consequently, if the engine is equipped in this manner, the surplus steam will blow-off in spite of all the driver's efforts to prevent it. To comply with this piece of Parliamentary wisdom it is necessary to retain the steam in the boiler, no matter how excessive the pressure may have unluckily risen, thereby placing the lives of human beings on and around the boiler in the greatest peril from an explosion, so that horses shall not be frightened by the steam issuing from the safety valve. How can we sufficiently censure this foolish regulation, which, to put it in other words, causes the lives of men to be daily jeopardised by the safety valves of road locomotives being wedged down to a dangerous degree, so that a pair of wild animals shall not be made afraid ?

The second clause in the Acts which we shall notice is the one referring to smoke consumption :—" Every locomotive used on any turnpike road or highway shall be constructed on the principle of consuming its own smoke ; and any person using any locomotive not so constructed, or not consuming, as far as practicable, its own smoke, shall be liable to a penalty not exceeding £5," etc. Hundreds of summonses have been taken out against traction engine owners for not doing what this clause enjoins them to do, but what is at present impossible for them to do. If the law-makers had enacted that coke be mixed with the coal used on road locomotives, so as to prevent, as far as possible, the formation of smoke, it would have shown greater wisdom than to talk about consuming smoke in such small boilers, wherein the means used for preventing smoke cannot be applied. We may here remark that this smoke nuisance is, like the steam blowing off from the safety valves, mostly caused by the road locomotives having to be stopped for the convenience of vehicles passing. When the engine is steaming along the highway very little smoke is emitted if the fire is carefully stoked, and the damper door nicely manipulated, because the exhaust steam causes a quick draught, making the fire burn brightly, and little smoke is seen issuing from the funnel; but immediately the engine is stopped, thick black smoke pours out from the chimney, and the fire burns sluggishly because of the loss of the artificial draught. And it is mostly when the engine is stopped that convictions are brought against the drivers or owners for not consuming their smoke, these summonses being not unfrequently taken out by the very persons for whose convenience the engine was stopped.

It is exceedingly unfair, if not intolerably unjust, that the poor traction engine proprietor should be fined for not doing what is practically impossible, while the railway locomotive may make any quantity of smoke in any neighbourhood without running any risk of a conviction.

We now come to the clause which states that damage caused by road locomotives to bridges is to be made good by owners, and furthermore "that it shall not be lawful for the owner or driver of any locomotive to drive it over any bridge on which a conspicuous notice has been placed by the authorities liable to the repair of the bridge, that the bridge is insufficient to carry weights beyond the ordinary traffic of the district." The traction engine owner suffers great inconvenience, and is made to waste much valuable time by his machinery having to go miles out of its way in order to avoid some rickety bridge, on which a board has been fixed prohibiting road locomotives crossing.

Many of our bridges have been in use for half-a-century or more; they are allowed to fall out of repair, not a penny being spent upon their maintenance by the authorities, and as the law now stands it appears to encourage bridge owners to allow these structures to become the worse for wear; nay, they are rewarded for their neglect in allowing them to become unsafe. After the bridge has been declared insufficient to carry the weight of traction engines, this species of legitimate traffic may be prohibited from crossing, to the immense inconvenience of the trade and commerce of the district; moreover, should any damage be done to a bridge which is considered safe, the traction engine owner is liable to be saddled with the whole expense of the repairs, irrespective of the wear and tear caused by the regular traffic which has continually been going on since the bridge's erection. Because the traction engine happened to prove the last straw on the camel's back, it appears to us to be very unjust to levy the whole of the cost of the repairs on the road locomotive owner, and allow all the other traffic to pass over the bridge without contributing anything towards its maintenance.*

* Section 7 of the Act of 1861 Mr. Aveling says:—"Renders the owners of road locomotives liable for all damage, directly or indirectly caused by the breaking down of any bridge by the passage of such a locomotive. This liability does not attach to carriers of heavy weights when the weight is not drawn by steam." Mr. Aveling said, "This anomaly exists at the present time, that if

Instead of prohibiting traction engines from crossing bridges, we think the bridge authorities should be compelled to make them strong enough to bear with ease the weight of all legitimate traffic, including the much abused road engine, which must now be considered as part of the ordinary traffic of every turnpike road and bridge.

Perhaps the most harassing annoyance that traction engine owners have to submit to is the restriction of the hours in which these engines are allowed to travel in various districts. "County authorities, Town Councils of any borough which has a separate Court of Quarter Sessions, may make bye-laws regulating the use of traction engines on highways—prohibiting their use upon roads where they are satisfied such use would be attended with danger to the public, and restricting the hours during which they may be used on any public road, such restriction not to exceed eight consecutive hours out of the twenty-four." These local authorities have the power given to them, according to the present Acts, of interfering with a lawful trade, thereby causing road locomotive owners to waste all the valuable daylight hours, and compel them to work in the night. Notwithstanding the irritating nature of such a bye-law it has, however, been very unwisely enforced in numerous instances.

It is impossible to over-estimate the increased difficulties, dangers, and the excessive cost of working traction engines in the night. Many accidents have occurred owing to shortness of water, and lack of superintendence, while the men have been struggling against great odds and groping about in the dark with lanterns. Some of these accidents have ended fatally, and are clearly chargeable to the one-sided bye-laws

I send a boiler weighing 15 tons drawn by 15 horses over a country bridge and that boiler breaks the bridge I have nothing to pay, but if I send the same boiler over the bridge drawn by an engine weighing 8 tons, and that boiler breaks through the bridge, I have the whole expenses to pay."

enforced by our local authorities. The terrible boiler explosion which took place at Maidstone a few years ago, which will be remembered by every reader, was doubtless the result of the oppressive and short-sighted regulations in force in that district. We are quite certain that it will take traction engine drivers all their time to keep matters going on smoothly during daylight; and to be compelled by law to engage in such an occupation during the night is a disgraceful piece of injustice, and our only wonder is that such legislation is tolerated by traction engine owners and manufacturers.

This industry should not be hampered and hindered by a meddlesome party who are opposed to all progress, who only study their own interests and comfort, who call traction engines nuisances, because the beasts they bestride do not like to meet them. Our best interests ought not to be sacrificed, and part of our trade should not be crushed, to suit the whims of timid old gentlemen who have a weakness for harnessing wild creatures, which have really no right in our streets until they have been tamed.

In case the traction engine owner has fortunately escaped being fined for breaking any of the laws we have already referred to, he is certain, sooner or later, to be called upon to pay a most exorbitant sum levied by the road authorities, for damage alleged to be done to the highway by the wheels of his engine and train, as the present Acts "give the road authorities the power to recover expenses caused by excessive weight or extraordinary traffic." Now, it can be proved to demonstration, that the wheels of modern road locomotives do not cut up good roads so much as horses' hoofs do; furthermore, if the roads are well made and kept in good repair, these engines and their trains tend rather to improve them, than to do them any damage; but where the roads are rotten or weak, as they invariably are in many out-of-the-way districts, any kind of traffic will cut them up; timber carting and brick loading by horses being particularly notorious for cutting up bad roads, and this traffic cannot be said to improve good roads; but these teams, no matter how incessantly they pass along the highways, and no matter how deeply the ruts may be cut thereby, are neither termed 'excessive weights' nor 'extraordinary traffic,' consequently no notice is taken of the damage done; but let a traction engine steam over a badly kept road, and leave its footprints on the road's soft surface, the owner is at once arrested by the surveyor of the district, and the highway authorities, backed by the clause referred to, claim extraordinary expenses for damage incurred. Cases could be enumerated wherein the road authorities have claimed costs out of all proportion to the damage done. Seeing that no damage can be done to properly constructed roads by the locomotive and its team, providing that the engine does not weigh more than ten tons, and is fitted with driving wheels about 5ft. 6in. to 6ft. 6in. diameter, and 16in wide, and the trucks carrying not more than six tons each, are mounted upon wheels having tyres nine inches wide, and so constructed that the front and hind wheels do not travel in the same track. we fail to see how expenses can be fairly claimed by the road authorities because of excessive weight. The wheels of the train are much wider in proportion than are the wheels of an ordinary farmer's waggon, carrying two or three tons, and drawn by three horses.

We may here remark that spring mounted traction engines do much less damage to all kinds of roads than the springless ones do, and were all road locomotives mounted upon springs, or provided with spring wheels, this provision would save owners much expense and trouble.

The Sheffield memorial presented to the Local Government Board in 1882, after showing the damage and inconvenience caused by road engines in certain districts, said that in some neighbourhoods where the roads are under 21ft. wide, the traction engine seriously interfered with the ordinary traffic. At a meeting of the local dignitaries held at Maidstone some time ago, it was stated that some of their roads were so narrow "that it was impossible for a carriage to pass an engine, or to turn back." *The Engineer*, referring to this childish meeting said :—" If the roads are so narrow as this, one would think that inconvenience must also be felt when a large waggon loaded with straw, or when a horse rake, or a drill, most of them wider than an engine, has to be passed. These, however, are not steam engines, and the magistrates do not shy at them. Instead of trying to make the traffic accommodate or reduce itself to these narrow roads, it would be much more sensible and better to seek to get the roads widened to suit the requirements of modern traffic.

THE END.

The "PENBERTHY" AUTOMATIC PATENT INJECTOR,

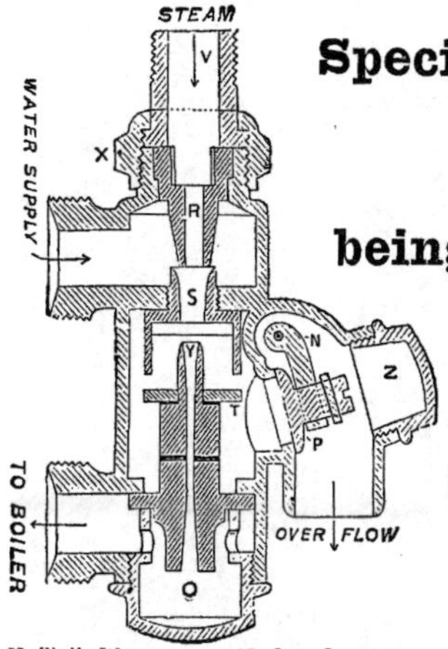

Specially adapted for use on Traction Engines, being absolutely automatic and re-starting.

Simple, Reliable, Cheap and Durable.

V Tail Pipe
Y Delivery Pipe
X Tail Nut
R Steam Jet
S Suction Jet
N Overflow Hinge
P Overflow Valve
T Ring
O Plug
Z Overflow Cap

All Jets interchangeable

SALE EXCEEDING 38,000.

THE 'XL' ALL EJECTOR
OR STEAM JET PUMP.

WORKING PARTS INTERCHANGEABLE.

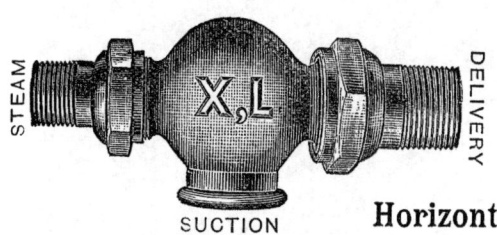

Reliable and Economical

May be placed Horizontally or Vertically.

PONTIFEX & WOOD, L$^{d.}$, Shoe Lane, London, E.C.

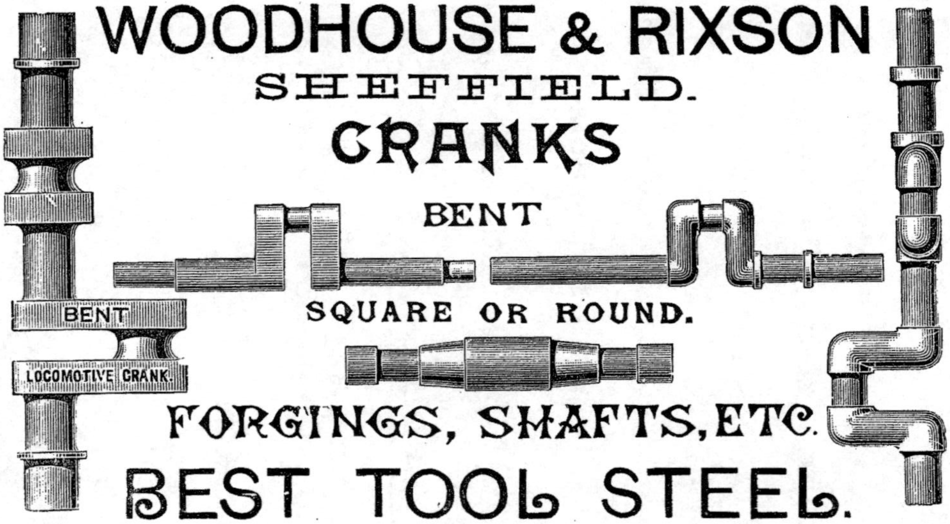

WOODHOUSE & RIXSON
SHEFFIELD.
CRANKS

BENT

SQUARE OR ROUND.

BENT LOCOMOTIVE CRANK.

FORGINGS, SHAFTS, ETC.
BEST TOOL STEEL.

BENT CRANKS MADE BY OUR IMPROVED METHOD possess the following advantages:—

1st.—The Rectangular shape of slotted-out Cranks is obtained, including the sharp angles of the Webs.
2nd.—The greatest amount of Hammering is effected where MOST NEEDED, *i.e.*, in the WEBS or CHEEKS.
3rd.—The FIBRE of the material is continued AROUND the Throw, and is not cut through transversely.
4th.—Each Throw is made in CORRECT POSITION, and not side by side, and WRENCHED into position by Twisting.
5th.—The expense of Slotting and Drilling is entirely avoided, and in many cases, Planing also.

DIRECTORY

OF

LEADING MAKERS

OF

ROAD LOCOMOTIVES & TRACTION ENGINES.

CONTENTS.

	PAGE.
W. ALLCHIN, Globe Works, NORTHAMPTON - -	299
AVELING & PORTER, ROCHESTER - -	292, 293
CHAS. BURRELL & SONS, THETFORD -	294, 295
E. FODEN, SONS & CO., SANDBACH -	296, 297, 298
R. HORNSBY & SONS, GRANTHAM - -	300
J. & H. McLAREN, Midland Works, LEEDS - -	301
RANSOMES, SIMS & JEFFERIES, IPSWICH -	302
ROBEY & CO., Globe Works, LINCOLN - -	303

292

AGRICULTURAL LOCOMOTIVES OR TRACTION ENGINES.

AVELING & PORTER,
ROCHESTER, ENGLAND.

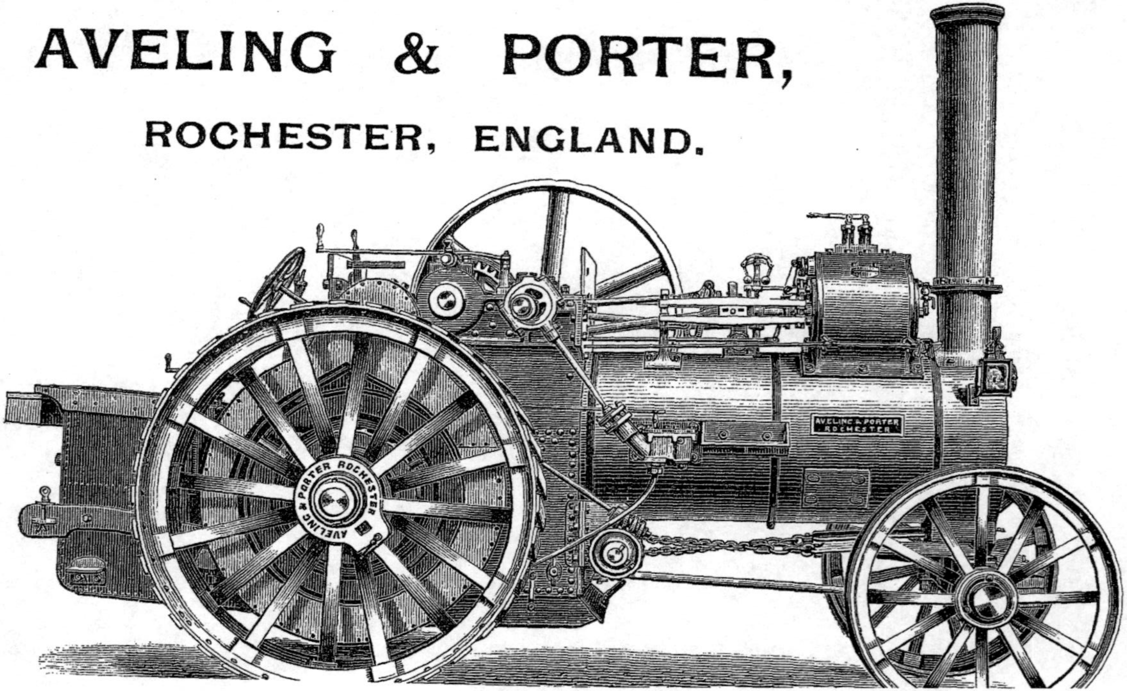

STEAM ROAD ROLLERS.

AVELING & PORTER,
ROCHESTER, ENGLAND.

MAKERS OF ROAD LOCOMOTIVES, MILITARY LOCOMOTIVES, CRANE LOCOMOTIVES, &c., &c.

CHARLES BURRELL & SONS, Lt

PATENT
SPRING MOUNTED ROAD LOCOMOTIVE

WE have now 60 Engines at work fitted with our Patent Spri arrangement, giving most excellent results, and the demand them is steadily increasing, since users of Traction Engines beginning to appreciate a saving of **50 per cent.** in the w and tear of the Engine which this system effects

This arrangement is considered by all who have seen it, a ridden on the Engines thus fitted, to be the only practical solut of the difficulty which has hitherto existed, of mounting Gea Traction Engines upon springs **without additional complicati** Since it does not increase the number of wearing parts, and th are no parts either to wear out, nor does it alter the design of Engine in any way.

St. Nicholas Works, THETFORD, Norfol

CHARLES BURRELL & SONS, Ltd.

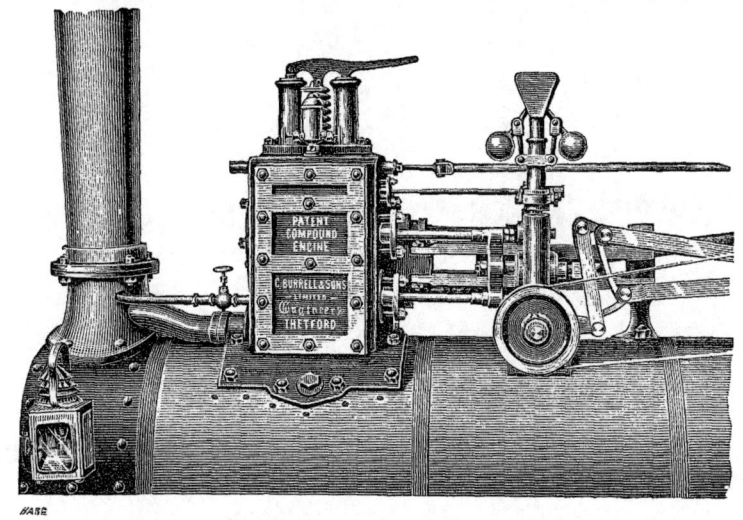

NEW PATENT SINGLE-CRANK COMPOUND SYSTEM.

This system has proved an **immense success**, and we have now a number of Engines at work upon this plan, both for Heavy Road Haulage and Thrashing purposes, giving extraordinary results in economy of fuel, and a marked absence of wear and tear. They are remarkable for the ease with which they start their load, and for great power developed.

FOR { **STRENGTH, SIMPLICITY, ECONOMY, DURABILITY, EFFICIENCY** }

This System has NO EQUAL, and possesses NO DISADVANTAGES.

Copies of Testimonials from users of these Compound Engines, testifying to the above results, on application to

St. Nicholas Works, THETFORD, Norfolk.

E. FODEN, SONS & CO., L:TD

SANDBACH,

Makers of all kinds of LAND & MARINE ENGINES from 1 to 1000 Horse Power.

SPECIALITY—

TRACTION ENGINES, Single or Double Cylinders, High Pressure, Twin or Tandem Compounds, with Slide or Piston Valves.

FACTS CONCERNING ECONOMY.

COAL.

The coal consumption of 1·84 lbs. per brake H.P. per hour accomplished by one of our Twin Compounds at the R.A.S.E. Trials, 1887, still stands the best on record for Traction Engines.

WATER.

Water consumption by the same Engine was 18·23 lbs. per brake H.P. per hour.

This Engine fitted with our Water-heater will travel 15 miles, exerting 18 I.H.P. without taking in a fresh supply.

Evaporation 12·96 lbs. of water per lb. of coal.

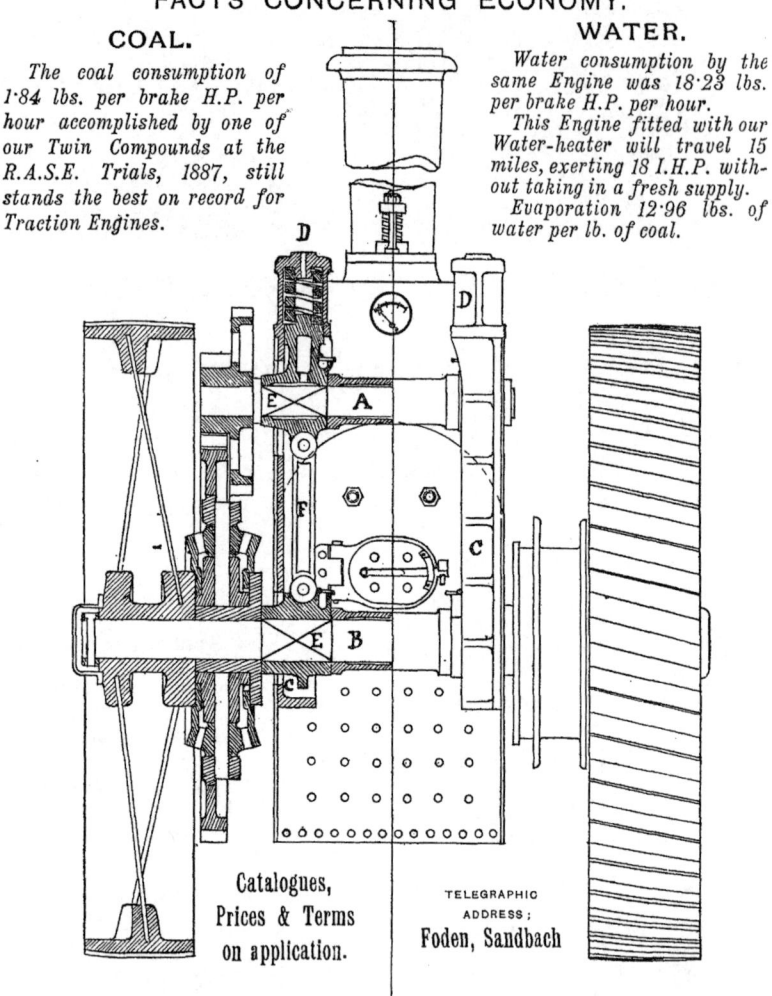

Catalogues, Prices & Terms on application.

TELEGRAPHIC ADDRESS:
Foden, Sandbach

E. FODEN, SONS & CO., LTD.

SPRING ARRANGEMENT

FOR

TRACTION & TRAMWAY ENGINES.

Since Traction Engines have become so indispensable as a means of conveyance, it is admitted that the greatest drawback to their more extended use, was the want of some means to obviate the excessive vibration whilst travelling. Most makers have endeavoured to overcome the evil by the use of india-rubber tyres, flexible spokes, spring wheels, and various methods of suspending or placing the weight of the Engine on Springs.

Flexible spokes were tried by us in 1880. They undoubtedly saved the Engine from vibration, but fatal defects soon became apparent and we discarded them (before any other makers had tried them) and, after carefully studying this matter and fully experimenting thereon, we found, and have since proved, that the arrangement as shown on the previous page (patented in 1882) fully meets all the requirements of the case, and has since then been applied to every Traction Engine made by us, with unqualified success, neither breakage or renewals to any of the springs having yet taken place.

DESCRIPTION.—On the preceding page is a sectional engraving, illustrating our Patent Spring Arrangement. It will be seen that the vibrating shafts are so arranged as not to alter their relative distance, at the same time allowing the weight of the engine to be carried by the springs. The bearings of the main axle and 3rd motion shaft A and B are connected by 2 levers F, the whole sliding in two axle boxes C C., preparation being made on the top of the two upper bearings for the reception of two strong coil springs contained in the cylinders D.

The bearings E E of the main axle and third motion shaft are of extra length, and parallel, and being coupled by the levers F, having joints at either end, the necessary oscillating or vertical motion is allowed to take place without locking or strain.

The hitherto great difficulty of accommodating the gearing on the stationary to the moving shafts is overcome in a very simple and effectual manner. The third motion shaft, which moves, is fixed slightly below a horizontal line drawn through the centre of the second motion shaft, which is stationary, so that the vibration— which at the most is only half an inch, $i.e.$, ¼-in. on either side of the centre line— does not practically alter the depth in gear of the two wheels.

There is also a similar spring on the fore axle contained in the cylinder, on which are cast stops to prevent the fore wheels coming in contact with the barrel of the boiler.

This perfect Spring Arrangement materially reduces the effect of shocks or vibrations caused by passing over rough roads, and is conducive to the reduction of the wear and tear arising from such causes in ordinary Traction Engines, as leaky fire boxes, tubes and joints, strained frames, and the jolting to pieces of the motion work throughout, and adds very considerably to the comfort of the driver.

REPORT OF THE ROYAL AGRICULTURAL SOCIETY ON E. FODEN, SONS & CO.'S, LTD.

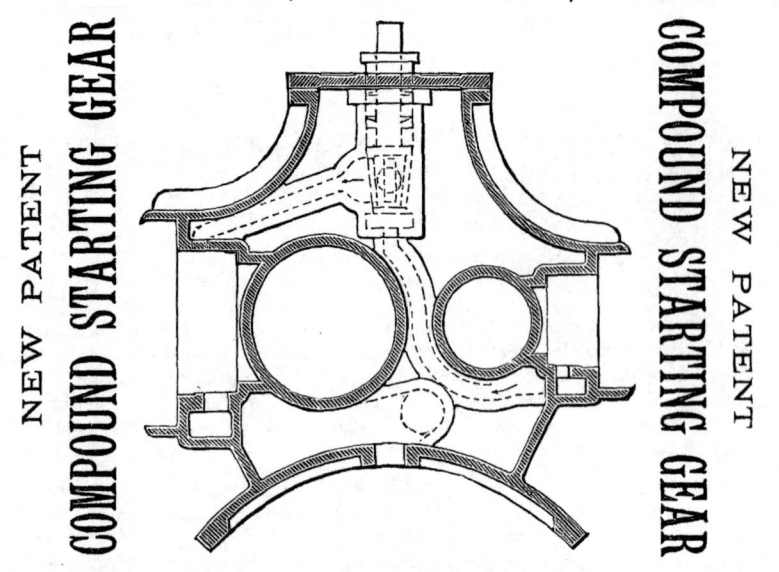

NEW PATENT COMPOUND STARTING GEAR

The "Compound Engine" is provided with a special arrangement by which the compound action may be instantly suspended, and both cylinders may take high-pressure steam, exhausting directly and independently into the funnel, the steam being supplied in such a manner that each cylinder shall give off the same amount of power. The object of such an arrangement is to give increased power to the engine when starting or doing exceptionally heavy work, as on steep gradients or when getting over soft ground. It is effected in the following manner:—In the passage between the high and low-pressure cylinders, a three-way cock is fitted, this cock being actuated either by an independent lever or else by the starting lever. In ordinary work, the steam from the high-pressure cylinder passes into the larger (or low-pressure) cylinder, is there further expanded, and exhausts therefrom into the funnel.

If, in case of emergency, it is required to get more power out of the engine, the above-mentioned three-way cock is opened, so that the exhaust from the high-pressure cylinder passes direct into the chimney, which relieves that cylinder of the back-pressure due to working the low-pressure cylinder, and consequently increases its power. Live steam is at the same time admitted into the low-pressure cylinder; but as this cylinder is so much larger than the high-pressure one, it is obvious that if steam of equal pressure were admitted to both cylinders, the larger one would do the most work, and consequently the engine would run unevenly. To overcome this, a steam reducing valve is provided in the passage to the low-pressure cylinder, by means of which the power to each cylinder is equalised, and the engine works as an ordinary double high-pressure engine.

The advantages of this starting gear are: 1st—It enables the user to obtain a great amount of power for starting purposes, getting out of soft places, or taking heavy loads up steep gradients. 2nd—In case of accident to either engine the one may be used independently of the other; for instance, supposing an eccentric rod broke on either engine, all the driver would have to do would be to uncouple both eccentric clips, set the slide valve of the disabled engine in the centre of its stroke, and open the three-way cock, as for working double high-pressure; by so doing the engine can be run as a single high pressure engine, until such time as it can be repaired. Thus by means of the starting gear, our Compound may be converted into double or single high-pressure engines; the additional complications being only the three-way cock, and the lever for actuating the same.

ESTABLISHED 1847.

ECONOMICAL TRACTION ENGINE.

WILLIAM ALLCHIN,
GLOBE WORKS, NORTHAMPTON.

ESTABLISHED 1847.

R. HORNSBY & SONS, Limited

HORNSBY'S IMPROVED NARROW-GAUGE TRACTION ENGINE
Sizes made:—5, 6, 8, and 10 Horse Power.

HORNSBY'S ROAD LOCOMOTIVE-HAULING THRASHING MACHINE

HORNSBY'S AGRICULTURAL LOCOMOTIVE ENGINE, with Four-Furrow Plough—Direct Traction System.

Manufacturers of IMPROVED STEAM ENGINES, from 1 to 500 Indicated Horse-power, suitable for Manufacturing, Electric Lighting, Mining and General Purposes.
Also AGRICULTURAL MACHINERY, including—Sheaf-Binding Harvesters, Mowing and Reaping Machines, and all kinds of Ploughs for General and Special Purposes.

COMPLETE CATALOGUES POST FREE.

Spittlegate Iron Works,　　　　84, Lombard Street,
GRANTHAM.　　　　　　　　　LONDON.

McLAREN'S TRACTION ENGINES

EFFICIENCY, ECONOMY, SIMPLICITY.

ARE DISTINGUISHED FOR

Made in all sizes from 6 H.P. upwards.

Specially adapted for
AGRICULTURISTS
CONTRACTORS
MILLERS
BRICKMAKERS
QUARRY OWNERS
ENGINEERS
&c., &c.

SPECIAL ROAD ENGINES FOR THE CONVEYANCE OF PASSENGERS OR LIGHT GOODS.
COMPOUND TRACTION, PORTABLE AND SEMI-FIXED ENGINES.
TRACTION WAGONS, STEAM OMNIBUSES, SLEEPING VANS, &c., &c.

McLAREN'S STEAM ROAD ROLLERS, STEAM PLOUGHING ENGINES AND PORTABLE ENGINES

are the best in the trade.

McLaren & Boulton's Patent BLOCK WHEELS

Enormously increase the Tractive Power of Road Engines.

An 8 H.P. Engine fitted with these wheels moved a 30 ton Boiler from Dukinfield to Oldham without difficulty, or damage to the roads.

Catalogues and Full Particulars on Application to

J. & H. McLAREN, Midland Engine Works, LEEDS.

RANSOMES, SIMS & JEFFERIES, L^{D.}

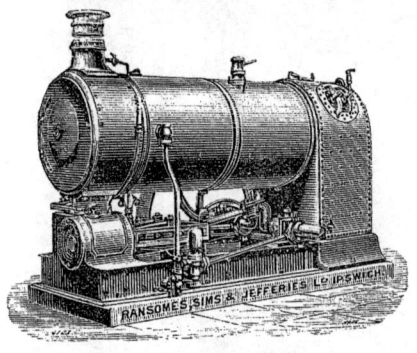

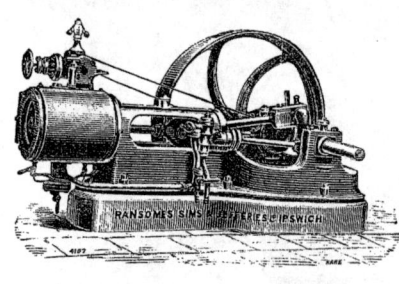

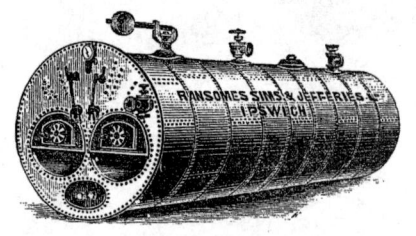

Engineers
AND
Boiler Makers

MANUFACTURERS OF

SIMPLE & COMPOUND
TRACTION ENGINES
PORTABLE ENGINES
SEMI-PORTABLE ENGINES
UNDERTYPE ENGINES
SHORT-STROKE ENGINES
LONG-STROKE ENGINES
GIRDER-FRAME ENGINES

CORNISH BOILERS
LANCASHIRE BOILERS
VERTICAL BOILERS
MULTITUBULAR BOILERS,
ETC., ETC.

ELECTRIC LIGHT
ENGINES,
WINDING, HOISTING,
AND
PUMPING ENGINES.

RANSOMES, SIMS & JEFFERIES' Factory, the ORWELL WORKS, established in 1789, is situated at Ipswich, 69 miles from London, and accessible by the Gt. Eastern Railway in an hour and a half.

It covers more than 12 acres of land and employs upwards of 1400 men and boys.

Full Illustrated Catalogues free by post on application.

ORWELL WORKS, IPSWICH.

ROBEY & CO., GLOBE WORKS, LINCOLN.

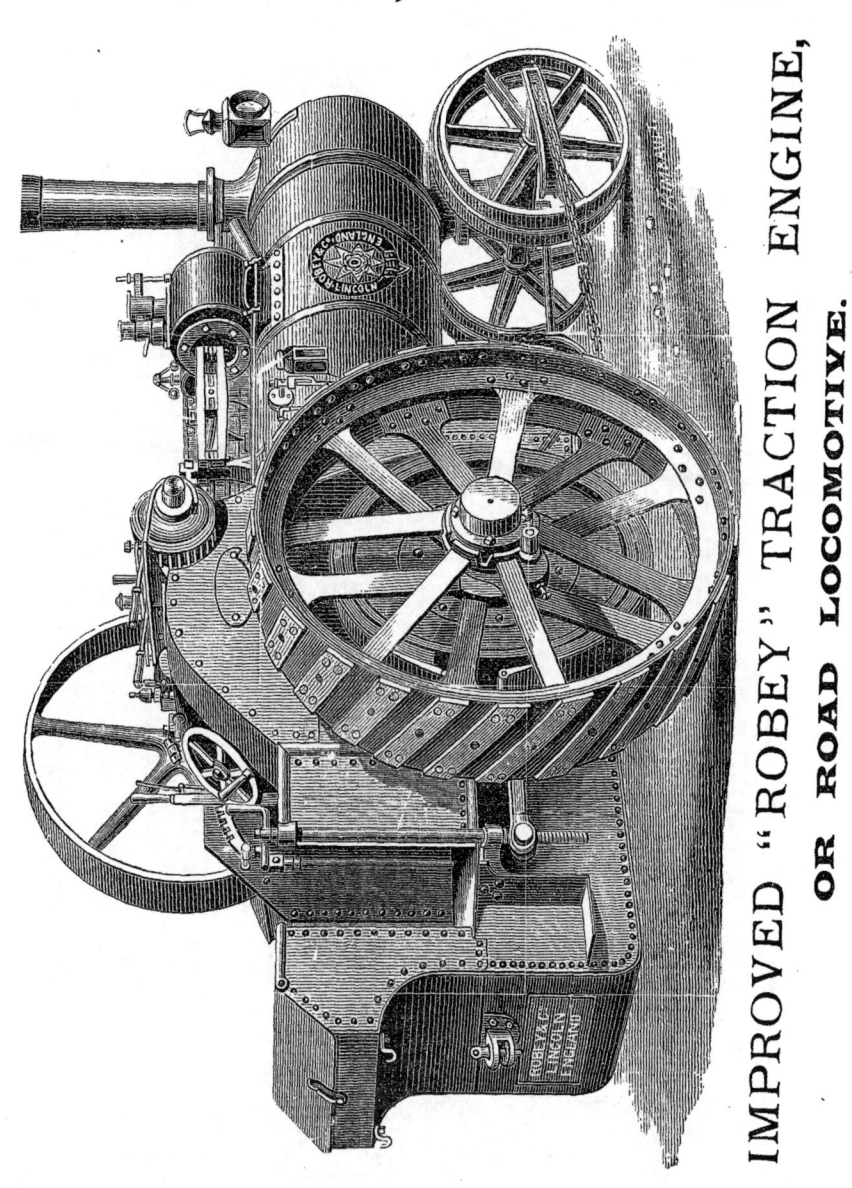

IMPROVED "ROBEY" TRACTION ENGINE, OR ROAD LOCOMOTIVE.

ROBEY & CO., GLOBE WORKS, LINCOLN.

INDIA RUBBER TYRES,

CONTINUOUS AND SEGMENTAL,

For the Wheels of ROAD STEAMERS.

MANUFACTURED BY

The NORTH BRITISH RUBBER CO., Limited,

Castle Mills, EDINBURGH.

Original Manufacturers of RUBBER TYRES, (continuous and segmental) for the THOMSON and other TRACTION ENGINES.

ALSO

Manufacturers of BELTING, HOSE, VALVES, RINGS, PACKING and other Articles for Mechanical and Technical uses.

Warehouses in LONDON, MANCHESTER, LIVERPOOL, LEEDS, NEWCASTLE, GLASGOW, & EDINBURGH.

Telegrams 'FLETCHER BROS., GRIMSBY.'

FLETCHER BROS. & CO.,

Manufacturing Chemists,

GRIMSBY.

(Contractors to Her Majesty's Government.)

SEND FOR PRICE LIST.

FLETCHER'S PINO-PHENOL PURIFIER

Concentrated Non-Poisonous, Non-Corrosive Disinfectant, Fluid or Powder.

USED BY

HER MAJESTY'S GOVERNMENT.

TOWN COUNCILS.   HOSPITALS AND ROYAL INFIRMARIES.
STEAM SHIP COMPANIES.   SCHOOL BOARDS.

Specially Prepared for and Used by all Well-Managed Engineering Works,
AND SANCTIONED BY HER MAJESTY'S BOARD OF TRADE.

Patent ELECTRICAL TACHOMETER,
OR
SPEED INDICATOR

(For full description see "ENGINEERING," Dec. 3rd, 1886.)

IN USE AT WOOLWICH ARSENAL.

STANDARD TYPE, FOR STATIONARY USE.

VERTICAL TORPEDO BOAT TYPE.

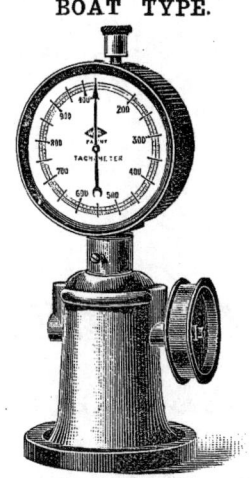

Price, complete with Pulley, £7 10s.   Price, complete with Pulley, £8 10s.

Portable Type in Elegant Morocco Case, £5 5s.

THESE INSTRUMENTS HAVE BEEN SUPPLIED TO—

Royal Carriage Department, Woolwich Arsenal
Royal Gun Factories, Woolwich Arsenal
Royal Laboratory, Woolwich Arsenal
Works Department, Woolwich Arsenal
The Admiralty, Portsmouth
Paterson & Cooper, Pownall Road, Dalston
W. H. Preece, Esq., G.P.O.
C. A. Wells, Etna Iron Works, Lewes
Crompton & Co., Ltd., Arc Works, Chelmsford
Easton & Anderson, Erith
Platt Bros., Oldham
Fry, Miers & Co., London
G. Spagnoletti, Esq., G.W.R.
Fyfe Main Electric Light Company, Brixton
Johnson & Phillips, London
John Collins, Denny, N.B.
F. Browne, 37, Lombard Street
Bryan, Donkin & Company, Southwark Park Road
North London Railway Company, Bow

"The Engineer" Printing Works
"Daily Telegraph" Printing Works
"Daily News" Printing Works
"Standard" Printing Works
Herman Kuhne, 25, New Broad Street
Lawrence & Co., Rice Mills, Rotherhithe
Marshall, Sons & Co., Gainsboro'
Peter Brotherhood, Lambeth
Aylesbury Dairy Company
Siemens Brothers, Charlton
Reid Brothers, Wharf Road, City Road.
Capt. Card.w's School of Military Engineering, Chatham
Maxwell, Butcher & Stevens, 6, Wormwood Street, E.C.
India-Rubber, Gutta-Percha & Telegraph Works Company, Silvertown.
Jas. C. Amos, West Barnet Lodge, New Barnet
Sir F. W. Truscott, Oakleigh, East Grinstead, Sussex
H. Hughes & Son, Fenchurch Street

&c., &c.

DETAILED ILLUSTRATED PRICE LIST ON APPLICATION TO

WOODHOUSE & RAWSON
UNITED, LIMITED,

Engineers and Electrical Contractors,

88, Queen Victoria Street, LONDON, E.C.

THE
PHOSPHOR BRONZE CO.
LIMITED,

LONDON, BIRMINGHAM, LIVERPOOL

AND ETRURIA.

Sole Makers of the following SPECIALITIES:—

PHOSPHOR BRONZE, "Cog Wheel" and "Vulcan" Brands.

"DURO METAL" *(Registered Title).* For Roll Bearings Wagon Brasses, &c.

PHOSPHOR TIN, "Cog Wheel" Brand. The best made

PLASTIC METAL, "Cog Wheel" Brand. The best in the market.

"PHOSPHOR" WHITE BRASS. Qualities I. & II

BABBITT'S METAL. "Vulcan" Brand. Qualities I. II., III., IV.

WEILLER'S PATENT SILICIUM BRONZE ELECTRICAL WIRE, For Overhead Lines.

Please apply for Circulars containing full particulars to the

COMPANY'S HEAD OFFICE,
87, Sumner Street, Southwark, London, S.E